PRAISE

INVISIBLE G

"In the rich tapestry of American hi country's first two Black generals remains a lesser-known gem, hidden in the shadows of time. These trailblazers, courageous leaders, and unyielding advocates rose above adversity with the unwavering spirit of justice burning in their hearts. Their story serves as an enduring beacon of hope, a reminder that the path to equality and justice is not without challenges, but it is one worth traversing."

—Ben Crump, Black America's Attorney General

"Documenting unsparingly the opposition they had to overcome due solely to the color of their skin, Melville recounts [his family's] stories with deep emotion, sympathetic with their struggles and angry with the system that made these humans invisible."

—*Booklist*

"The book Black America needs in this moment. This American story is rooted in family, loyalty, heart, and excellence. Doug's family is representative of our own families . . . rooted in the consistent legacies of connection, disruption, and love in ways we can all be proud of."

—Eboni K. Williams, journalist, attorney, and author of *Pretty Powerful*

"Melville traces his family's history to ensure their groundbreaking achievements are not forgotten."

—*Arlington Magazine*

"For centuries, Black people's contributions to American history have been overlooked—including their contributions to US military prowess. As Black Americans, our stories have power. And it's long past time to reclaim that power."

—Charlamagne Tha God, *New York Times* bestselling author of *Black Privilege*

"*Invisible Generals* is not Black history, nor military history, but American history. The military's first two Black generals paved the way for so many not just in our military but for all Americans—yet they have been invisible for so long. This book will educate and inspire Americans to honor these heroes for their selfless contributions toward moving our nation forward."

—Brigadier General (Ret.) Leon Johnson, former national president of Tuskegee Airmen, Inc.

"A thoughtful, highly readable blend of family and military history."

—*Kirkus Reviews*

INVISIBLE GENERALS

REDISCOVERING FAMILY LEGACY, AND A QUEST TO HONOR AMERICA'S FIRST BLACK GENERALS

DOUG MELVILLE

BLACK PRIVILEGE PUBLISHING

ATRIA

NEW YORK LONDON TORONTO SYDNEY NEW DELHI

ATRIA

An Imprint of Simon & Schuster, LLC
1230 Avenue of the Americas
New York, NY 10020

First Black Privilege Publishing/Atria Paperback edition December 2024

BLACK PRIVILEGE PUBLISHING / ATRIA PAPERBACK and colophon are trademarks of Simon & Schuster, LLC

Simon & Schuster: Celebrating 100 Years of Publishing in 2024

Interior design by Timothy Shaner, NightandDayDesign.biz

Manufactured in the United States of America

1 3 5 7 9 10 8 6 4 2

Library of Congress Control Number: 2023940772

ISBN 978-1-6680-0513-2
ISBN 978-1-6680-0514-9 (pbk)
ISBN 978-1-6680-0515-6 (ebook)

To my mom, Sonja D. Melville, and my dad,
Judge L. Scott Melville.

Thank you for believing in me and supporting me unconditionally in life and through love.

And to every parent who has sacrificed for a lifetime to manifest their child's dream.

CONTENTS

Three Centuries, Five Generations

LOUIS PATRICK HENRY DAVIS

Unknown–1930

Louis was a household servant. After the Civil War ended, he became one of the first Blacks to work for the federal government. He and his wife, Henrietta, had three children. Benjamin O. Davis was their youngest son.

BENJAMIN OLIVER DAVIS SR. "OLLIE"

1880–1970

Ollie became America's first Black general. He had three children with his wife, Elnora, who died in childbirth. Benjamin O. Davis Jr. was the middle child and only son.

BENJAMIN OLIVER DAVIS JR.
"BEN"

1912–2002

Ben was America's second Black general. He and his wife, Agatha, had no biological children. However, they raised Larry Melville, the son of Agatha's sister, Vivienne Melville, as their child.

LAWRENCE "L." SCOTT MELVILLE
"LARRY"

1933–

Larry, Ben and Agatha's nephew, moved in with them when he was around seven years old. He graduated from Howard University School of Law and became a Connecticut superior court judge. He and his wife, Sonja, had three children. Doug Melville is their youngest son.

DOUG MELVILLE

1977–

Doug is the author of *Invisible Generals* and a leader in the field of corporate diversity.

Timeline*

YEAR	EVENTS IN US HISTORY	EVENTS IN FAMILY HISTORY
Unknown		Louis Davis born
1861	Abraham Lincoln elected president; Civil War begins	
1865	Civil War ends; Reconstruction begins	
1866	Buffalo Soldiers regiment forms	
1873	President Ulysses S. Grant's second inauguration	
1875		Louis Davis marries Henrietta Stewart
1880		Benjamin O. Davis Sr. ("Ollie") born
1889	Charles Young graduates from West Point	
1893		Louis and Henrietta purchase home at 1830 11th Street NW in Washington, DC
1897	William McKinley becomes president	
1898		Ollie enlists in the Army
1901	William McKinley assassinated	Ollie promoted to second lieutenant of cavalry

* This timeline provides highlights from my family story and its intersection with American and world history. Many thanks to the National Air and Space Museum of the Smithsonian Institution, which compiled an incredible timeline of Ben's life, and from which some of these dates and events were taken.

YEAR	EVENTS IN US HISTORY	EVENTS IN FAMILY HISTORY
1902		Ollie marries Elnora Dickerson
1903	Wright brothers take first flight	
1905		Olive Elnora Davis born
1912		Benjamin O. Davis Jr. ("Ben") born
1914	World War I begins	
1915		Ollie reports as instructor at Wilberforce University, Ohio
1916		Elnora Dickerson Davis born; Ollie's wife, Elnora, dies
1917	US enters World War I	
		Ollie is deployed to the Philippines
1918	World War I ends	
1919		Ollie marries Sadie Overton
1920		Ollie's family moves to Tuskegee, Alabama; Ollie teaches at Tuskegee Institute
1924		Ollie's family moves to Cleveland, Ohio
1925	Army War College report issued	
1926		Ben takes his first plane ride and is determined to become a pilot
1930		Ben moves to Chicago; Louis Davis dies
1931		Ben fails his first attempt at West Point examination

YEAR	EVENTS IN US HISTORY	EVENTS IN FAMILY HISTORY
1932		Ben reports to West Point
1933	President Franklin D. Roosevelt inaugurated	Lawrence "L." Scott Melville ("Larry") born
1934		Ben begins dating Agatha Scott
1935		Ben's application for pilot training is rejected
1936		Ben graduates from West Point as a second lieutenant and marries Agatha
1937	World War II begins in Asia	Ben reports to Fort Benning, Georgia; Ben and Ollie begin traveling to Black colleges to train and teach ROTC students
1938		Ben reports to Tuskegee Institute as professor of military science
1939	World War II begins in Europe	Ben promoted to first lieutenant
1940		Ollie promoted to brigadier general; Ben temporarily promoted to captain; Larry moves to Tuskegee to live with Ben and Agatha
February 1941		Ben is assigned as Ollie's aide-de-camp at Fort Riley, Kansas
March 1941		Black volunteers solicited for pilot training in Tuskegee and the 332nd Fighter Group activated

YEAR	EVENTS IN US HISTORY	EVENTS IN FAMILY HISTORY
May 1941		Ben reports to flying school at Tuskegee Army Air Field
December 1941	Japanese forces attack Pearl Harbor, Hawaii; US enters World War II	
1942		Ben transferred to Army Air Corps; temporarily promoted to major, then lieutenant colonel*
August 1942		Ben assumes command of the 99th Fighter Squadron in Tuskegee
1943	99th transferred to Morocco, Tunisia, then Sicily	
September 1943		Ben returns to the US
October 1943		Ben assumes command of 332nd Fighter Group in Michigan
1944	332nd transferred to Italy	
		Ben temporarily promoted to colonel; at Ben's request, the tails of the Tuskegee Airmen's P-51 Mustangs are painted red
April 12, 1945	President Harry S. Truman inaugurated	
May 1945	World War II ends in Europe	
June 1945		Ben returns to the US
September 1945	World War II ends in Asia	
1946		Ben permanently promoted to captain and assumes command of Lockbourne AAB, Ohio

* Particularly during wartime, the US military sometimes used temporary promotions to allow qualified personnel to fill vacancies.

YEAR	EVENTS IN US HISTORY	EVENTS IN FAMILY HISTORY
1947	The Air Force (USAF) becomes a separate military branch	
April 1948	USAF announces integration	
July 2, 1948		Ben permanently promoted to lieutenant colonel
July 20, 1948		Ollie retires from the US Army
July 26, 1948	President Truman signs Executive Order 9981	
August 1949		Ben reports to Air War College
1950	Korean War begins	Ben permanently promoted to colonel; Henrietta Davis dies
1953	Korean War ends	
1954		Ben temporarily promoted to brigadier general and receives first star
1955	Vietnam War begins	
1959		Ben temporarily promoted to major general and receives second star; Larry begins studies at Howard University School of Law
1960		Ben permanently promoted to brigadier general
1961	President John F. Kennedy inaugurated; US becomes more deeply involved in Vietnam War	
1962		Ben permanently promoted to major general; Larry graduates from Howard law school and moves to Connecticut

YEAR	EVENTS IN US HISTORY	EVENTS IN FAMILY HISTORY
1963	President Lyndon B. Johnson sworn in after assassination of President John F. Kennedy	Larry marries Sonja Douglas
1964	Civil Rights Act passed; first Black commercial pilot hired by American Airlines	
1965	Voting Rights Act passed	Ben promoted to lieutenant general and receives third star and reports to Seoul, Korea; Sonja-Lisa Melville born
1966		Sadie Overton Davis dies; Scott D. Melville born
1967	Thurgood Marshall becomes first Black Supreme Court justice	
1969	President Richard Nixon inaugurated	
January 31, 1970		Ben retires from USAF as lieutenant general
February 1, 1970		Ben becomes director of public safety for Cleveland, Ohio
June 1970		Ben joins President's Commission on Campus Unrest
July 1970		Ben resigns from Cleveland position
September 1970		Ben becomes consultant for US secretary of transportation on airline security
November 4, 1970		Ben becomes director of civil aviation security for Department of Transportation

YEAR	EVENTS IN US HISTORY	EVENTS IN FAMILY HISTORY
November 26, 1970		Ollie dies
November 30, 1970		Ollie is buried at Arlington National Cemetery
1971		Ben becomes assistant secretary of transportation for safety and consumer affairs
1972	Tuskegee Airmen, Inc. founded	
1973	US withdraws from Vietnam	
1974	President Nixon resigns; Gerald Ford sworn in as president	
1975	Daniel "Chappy" James Jr. becomes first Black four-star general	Ben retires from civil service
1976	First female cadets admitted to USAF Academy	Ben works as special assistant to secretary of transportation for national maximum speed limit
1977	President James "Jimmy" Carter inaugurated	Ben joins President's Commission on Military Compensation; Larry becomes a judge; Doug Melville born
1981	President Ronald Reagan inaugurated	Ben leaves position as special assistant to the secretary of transportation
1986		Ben makes his final visit to the West Point campus

YEAR	EVENTS IN US HISTORY	EVENTS IN FAMILY HISTORY
1989	President George H. W. Bush inaugurated	
1991		Ben publishes his autobiography; awarded a lifetime membership in Tuskegee Airmen, Inc.
1993	President William "Bill" Clinton inaugurated	
1998		Ben promoted to four-star general (retired)
2000	Lieutenant Colonel Shawna Kimbrell becomes first Black female fighter pilot in USAF	
2001	President George W. Bush inaugurated	
March 10, 2002		Agatha dies
July 4, 2002		Ben dies
July 17, 2002		Ben buried at Arlington National Cemetery
2009	President Barack Obama inaugurated	
2011	*Red Tails* movie screening	
2012	*Red Tails* movie debuts in theaters	
2013		Larry retires from being a superior court judge
2015		Doug's first meeting at West Point to discuss Davis Barracks
January 20, 2017	President Donald Trump inaugurated	
August 12, 2017	"Unite the Right" rally in Charlottesville, Virginia	
August 18, 2017	Davis Barracks dedication at West Point	
2019	Davis Airfield naming ceremony at USAF Academy	

YEAR	EVENTS IN US HISTORY	EVENTS IN FAMILY HISTORY
2020	Colonel Charles McGee (retired) promoted to brigadier general	
		Google's commercial "The Most Searched: A Celebration of Black History," featuring Ben, airs during Grammy Awards; Aunt Jemima retired
2021	President Joseph "Joe" Biden inaugurated; Congress establishes naming commission to evaluate Confederate fixtures	
2022		Quarters 64 at F. E. Warren Air Force Base dedicated to Ollie
October 2022		Doug visits Ollie's archives at US Army Heritage Center and Ben's archives at Smithsonian's Air and Space Museum
December 17, 2022		Ben inducted into Paul E. Garber Shrine at the Wright Brothers National Memorial in Kitty Hawk, North Carolina
August 2023		Mural of Ben in front of his P-51 Mustang and Tuskegee Airmen memorial unveiled in Campomarino, Italy

PREFACE

A Man on a Mission

While my friends and peers were starting families of their own, I was on a mission to find mine.

Though my family wasn't broken or broken up, its history had been splintered into bits and pieces. And many of those pieces had been lost—some buried intentionally, some lodged only in people's memories, some fading with the passing of time.

Like many people do, I thought of family history as something that happened "back then." Events of days long since passed that have no bearing on the present. Yet that couldn't be further from the truth.

Whether we realize it or care to acknowledge it, beyond biologically, we're the receipts of our ancestors' journeys. The accomplishments of those behind us impact numerous aspects of our existence, from where we live and what school we attend to our friends and what jobs interest us. And for me and many other Americans, our ancestors' commitment to change the nation continues in us today.

When I began this adventure, I knew very little. But once the process started, I realized the story was even more robust and amazing than I could've imagined. I've spent over a decade leveraging the tools at my disposal—Google, eBay, LinkedIn, museums, historical documents, organizations and networks, my relatives' lived experiences—to grasp every strand I could find. I've woven these strands into a tapestry that spans three centuries and five generations, and I've seized every opportunity to take ownership of the narrative. I've looked at it from afar and up close, to reframe it and present it in a new way that's relevant for new audiences. It's a pursuit that has brought me—and still brings me—immense satisfaction and joy.

Along the way, I've learned much about myself and my identity as an American, as a man, and as someone from a mixed-race background. Most important, I've learned the high price some people have paid for the American dream—including Benjamin O. Davis Sr. and Benjamin O. Davis Jr., the Invisible Generals. Whether we're aware or not, each one of us has been impacted by a figurative invisible general—someone who worked behind the scenes to clear the path for those who follow.

As my great-uncle, General Benjamin O. Davis Jr., said, "The privileges of being an American belong to those brave enough to fight for them." In many ways, he and his father had to fight harder than most, as they blazed a trail to become the first two Black generals in the United States military. But they clung to the American dream and prioritized service to their family and the country. I truly feel that their sacrifices made the United States, and the world, a better place for all people.

People like the Invisible Generals lived by a singular vision: *You can use the system to defuse the system*. In their roles as military leaders, they couldn't take to the streets to protest. Instead, they utilized systems designed to hold them back, making American culture more inclusive by applying their insight and performance to the levers of power they had access to. The stories of these two men can show us different ways to transform the world with whatever power and tools we have access to.

In discovering my family's story, I've learned that some of the greatest lessons for advancing our lives forward are found in looking back and taking time to see the hurdles that have already been overcome, and to use past victories to fuel our own dreams of what's possible. Although extensive information about these two men already exists, this book documents my personal journey to rediscover the accomplishments of the Invisible Generals, both through the lens of America and the lens of my family. And I hope it encourages *you* to celebrate and be grateful for the heroes in your family and within your community, and elevate their stories for the world to hear and know.

Every generation is a continuation of the previous generation's evolution. By engaging with the past, we can add new dimensions to history—and make those who have been erased or overlooked either intentionally or accidentally become visible.

INTRODUCTION

What's in a Name?

The greatest story never told is right there in the living room—but no one is talking.

Who the hell is Colonel Bullard?

I glanced around the packed theater, straining to gauge the other viewers' reactions. Everyone else was completely captivated by the feature; *Red Tails*, George Lucas's movie based on the Tuskegee Airmen, the first Black fighter pilots in the US Army Air Corps.

Was it possible that no one else knew that the names of the actual pilots had been omitted from the film?

To be honest, I wasn't familiar with all the characters, either, but I definitely knew one: their commander, Benjamin O. Davis Jr. He'd raised my dad and been a kindly, generous grandfather figure in my life, as well as the centerpiece of my family. Bullard may have looked like Ben, but here his name was wrong.

It was complete happenstance that I'd even attended this screening. At the time, I was in full entrepreneur mode, and an event planning company I cofounded, RedCarpets.com, was peaking. As I boarded a plane to travel from New York City to Los Angeles for a big charity event we were red-carpeting at the London Hotel in West Hollywood, I received a call from my friend Nena. I fumbled with my carry-ons and answered the phone.

"Doug, what are you doing two days from now? I know this is really last minute, but I need you to fly to Dallas. There's going to be a *Red Tails* screening, and your family is in the movie. It's a private event, and Oprah will be attending the after-party. She's one of the producers or something," Nena said, her excitement palpable.

A few years earlier, Nena and I had connected while I was collaborating with her on a book project for Earvin "Magic" Johnson, whom I had been working for as part of his executive team. During our countless hours together, I'd tell Nena stories to lighten the mood. I *love* telling stories. Especially family stories. One of the family stories I'd told her was about General Benjamin O. Davis Jr., known to me as just Ben. The part of the story she liked most was when Ben received his fourth star at the White House. President Bill Clinton himself had pinned the star on Ben—and I was there in the front row. I couldn't believe she remembered this story well enough to recite it back.

Nena was coordinating the screening after-party, and she wanted me to attend as a representative of General Davis's family.

"We need you here. This is big," she said. "No one has contacted you about this?"

"No."

"Well, I am. Come tomorrow. Tell me you will be here."

"What movie is this again?" I asked.

"*Red Tails*, Dougie. It's George Lucas."

I vaguely remembered hearing something about the movie, but I didn't know much. If the cast was there, I thought that would be an amazing opportunity to meet the actor playing Ben.

My LA event was successful, and the next morning I woke up early, still unsure if it was worth flying to Dallas. I had refreshed the ticket prices, yet hadn't hit the purchase button. But I could hear Nena's voice in my head, and her excitement had gotten to me. I hadn't seen her in years, and I could leave LA a day early to go. So the next thing I knew, I was buying a one-way ticket from LA to Dallas.

When I boarded, I saw several of the film's actors on the flight. The plane took on a bit of a house-party vibe, with the flight attendants and pilots taking pictures and allowing guests to pose in the cockpit. It was evident that we were all headed to the same place. This was confirmed when I walked the aisle to introduce myself to semi-familiar faces. *Going to the* Red Tails *screening?* I'd ask. *Yep*, they'd respond. United in a common adventure, I felt connected to the actors. The energy built as we anticipated seeing the movie for the first time.

The plane touched down, and everyone separated. The actors' drivers and car services whisked them away, and Nena picked me up. It was an incredibly rainy day in December 2011. The windshield wipers swiped at full blast as we inched down the highway. We arrived at the theater later than planned, and upon entering I was surprised to see two reserved seats with my name on them. I'd be in the middle of the action, with the cast on one side and the actual Tuskegee Airmen on the other. I was a physical bridge between two realities: on my right, eighty-plus-year-old heroes, and on my left, the actors who played them in the movie. I could only imagine how the actors felt, sharing this moment with the men they portrayed.

As a young man who was ready to take on the world, I felt more aligned with the actors. They all looked good, fresh, and in their prime. The Airmen, on the other hand, were in their twilight years. In their faces you could see the effects of time and the long, hard road it had taken for them to get to Dallas.

The room was buzzing. Apparently, George Lucas had been working on the film for years, trying to get it produced, and people thought this might be the last film of his career, which ended up being true. It was surreal to see that one of the greatest moviemakers of our time had laid his hands on the story of the Tuskegee Airmen—and part of my family story.

The lights dimmed, and the president of the Tuskegee Airmen organization, General Leon Johnson, stood in front of the theater and said how exciting it was to see these American heroes on-screen. And it was an honor to share with us what we'd be watching. He was then joined by someone from the studio, who explained that we'd be viewing the most recent director's cut; the final version would debut in three months.

Okay, I'm not one hundred percent sure what that means, but they must be making more changes, I thought.

The silent theater was overtaken by the thundering rumble of a red-tailed P-51 Mustang, and we were off! I eagerly awaited Ben's debut on the big screen. I thought about the empty reserved seat next to me. I wished I had brought my dad with me. He would have *loved* this.

About halfway into the movie, one of the characters announced that the commander of the Tuskegee Airmen was coming. *This was it!*

And there he was, portrayed by Terrence Howard, all steely gaze and stern countenance—and he looked just like Ben.

This is going to be amazing!

But as the other characters greeted him, they referred to him as Bullard. *Colonel Bullard.*

Wait . . . Was this the right character? Who was this Bullard person? Where was Colonel Davis?

I was stunned.

After the movie ended, Nena found me and told me to hurry to her car so we could head to the after-party right away. She said she'd set up something special.

As the rain poured down around the car, all I could think was *Is that really the final cut?* Some of the special effects were still rough around the edges, so maybe they'd change the names, too? Would that be possible?

Walking into the after-party was like entering a who's who of Black Hollywood, everyone who was on the plane from LA plus all sorts of Black luminaries. Gayle King, David Oyelowo, Ne-Yo, and Nate Parker, among others, were all there. I briefly spotted Tyler Perry and T. D. Jakes before being whisked to a separate room for a private conversation with Oprah. The agency that had organized the event had arranged for her to have private interactions with some of the Airmen's families. This would be my first time meeting Oprah, who was at the height of her fame, and I looked forward to speaking with her. When she walked in, she seemed happy to meet me, since I was there as a representative of the general's family.[1]

After some friendly banter we started discussing *Red Tails.*

"So what do you think of this incredible movie everyone here has come together to make?" she asked.

Now, when you meet someone like Oprah for the first time, you'd think that every word would be permanently etched in your memory. However, beyond that first line, nothing registered because I was so fixated on the fact that someone *at this party* had changed Ben's name—as well as the other soldiers' names.

"It's great—but what about the names? All of the real names were omitted from the film," I said. I wasn't sure what I expected to hear.

"Douglas, this is Hollywood. This isn't a documentary. Each character is an amalgamation of three or four characters." And so forth.

As she explained all of this to me, over her shoulder I saw the group of about a dozen Tuskegee Airmen, lined up in their bright red jackets, worn to signify their Red Tails days. Most were in their eighties and nineties. Some were accompanied by nurses, spouses, or their children. Some towed oxygen tanks attached to breathing machines, and others sat in wheelchairs.

Later that evening, when I finally had a chance to speak to them, I heard that their living accommodations were quite meager, but they were grateful to travel to this event. That it was just nice to be thought of. Many of them, in one way or another, told me they were simply happy to be alive.

Maybe the open bar and jet lag were kicking in, but I felt I was living in a liminal space, between two realities. In a world where you'd invite someone to a movie that's about them—about the racism, discrimination, and trauma they endured in wartime and beyond—*and not even use their real names.* These men had fought to tell the story to honor and preserve their family names, to protect both their families and their country's freedom. To be recognized and respected. It wasn't about money—it was about duty, honor, and country.

As the evening continued and I met various cast members and others affiliated with the movie, I slipped the question into every conversation:

What about the names?

Even after hearing various explanations—we tried to; we had to do some reshoots; it was about making sure this story made it to life; we couldn't negotiate with every family, but they were brought on as advisors—I wondered if I was missing something.

As the night went on, and the champagne supply dried up, I remained disturbed. *Why would you make a movie like this and not give credit to the people who lived the story?* I internally began transitioning from Team Hollywood to Team Airmen, from a desire for entertainment to a desire for authenticity.

The busy day, which had started fifteen hours earlier, should have found me sleeping as soon as my head hit the pillow in my hotel room that night. But between riding high on the energy of the screening, experiencing the sensory overload of the after-party, and being so confused and uneasy about the name changes, I couldn't wind down.

The next morning I flew back to NYC early, and I made plans to continue on to Connecticut, to see my parents.

I was still riled up from the screening when I joined my dad at the kitchen table for breakfast, his favorite morning ritual. My dad, the Honorable L. Scott Melville, was a state superior court judge in Connecticut and he also happened to be the third Black judge to hold that position in the state.

"How was it?" he asked.

"Dad, they changed all the names. In the movie. They changed the names!"

I was in my feelings. After I finished my three-minute diatribe as he finished his first course of buttered toast, he began to smile, and then he began to laugh.

He laughed. Right in my face.

There I was, in my early thirties, running home to share the family outrage, and I'm getting laughed at.

"Well, Douglas, of course they changed the names."

"What are you talking about?" In my peripheral vision I could see my mother, Sonja, a lifelong educator, who had walked into the kitchen to watch this scene unfold.

"We're Black," he said.

The moment felt detached from time and space. I was speechless.

"Douglas, our names and our whole history as Americans has been changed because of race. Of course they changed the names."

My dad had never, ever said anything like this to me. He operated with a "say less" mentality, and this was the first time he'd said "Black" as though that was a burden. I couldn't wrap my mind around it.

Up until that point, I hadn't been so openly confronted with the idea that my family history could be adversely affected, or that my opportunities could be limited due to race. I was acutely aware that there was a racial construct in America. But my dad had taught me that education and being good to your family were critical, and that being Black was part of our American experience—we had to use what we have as a privilege. I operated as if being Black or mixed-race opened up opportunities—it was a superpower or advantage, not the other way around.

Yet now, here was my dad, eating his breakfast just like he did every day, matter-of-factly explaining to me that we're Black, and much of our entire history's been erased because of that one fact.

While I continued to sit there in stunned silence, he continued.

"I should tell you the story of the family, because I've never told you or your older brother and sister."

"The story of the family?"

"Yes. It goes back much further than me. I think it's important for you to know the story of our family."

"What?"

"Yeah, one day I'll tell you the whole story—or at least as much as I know." With that, he cleaned up his dishes, tossed his napkin aside, and stood. "But now, I have to go to work," he said and left.

I turned to my mother, still hovering in the corner of the room.

"Mom, what's all this about 'the story of the family'?"

"Dougie, why are you worried about that? Can't you worry about your business?"

"No, Mom—I want to know more about this."

"You always want to know about the family. It doesn't matter. What happened in the past happened in the past. You don't have any kids to pass it down to anyway." This wasn't the first time I'd heard a not-so-subtle jab from my mom about my single and childless life, and I doubted it would be the last. She then changed the subject back to the film screening.

"Did you get a picture of you and Oprah Winfrey? She is my favorite. I wanna show my friends."

When the final cut of *Red Tails* was released, my suspicions were confirmed: Ben's name, as well as those of the other Airmen, were still omitted.

Not long after that, my dad sat me down in the living room. He said he wanted to tell me his story, about how Ben Jr. had raised him and about why General Davis (which is how he referred to Ben's dad, Ben Sr., Ollie) was so influential on our family—and how silence and invisibility factored into our family story.

"Doug, the reason we've lived so well here in Connecticut and have everything we could ever ask for is because we don't make waves. We don't talk. We don't correct people. We just do the job. Ben never talked, and I never talked. And General Davis never talked. And there's nothing really written about us beyond the public records of our work. The history books are wrong and the photos are wrong and the words are wrong—everything's wrong. But we're still here, and we're alive. Our wealth is in our longevity."

I felt the same anticipation build as I'd felt at the screening, while waiting for Ben to appear on the big screen. And I'd be similarly disappointed because instead of telling me the story, my dad started to berate me.

"I think you'd be best served if you tried to get a master's degree or spent more time on your company. Why would you waste your time worrying about this foolishness? Nothing good's ever gonna come of it."

He went on from there, saying that if it were an interesting story, someone else would've already talked about it. Didn't HBO make a Tuskegee Airmen movie that had Ben as a character? Now there was this *Red Tails* movie too. Why was I complaining? How many movies did I want them to make?

I clearly wouldn't be hearing what I'd wanted to that day either.

But I persisted, and the next year my dad began to share our family history with me—what the generations before him had gone through, and what he himself had gone through. It gave me a perspective that helped me understand why having our names changed in a movie didn't register as an outrage for him as it had for me.

My dad told me the story of a father and son who became America's first Black generals. Trailblazing innovators, two of a kind, who'd not only fought for our country but also fought behind the scenes to chip away at America's hidden caste system, to provide more opportunities for all people. Yet, beyond the most fanatic military history buffs, almost no one has ever heard of them.

A Call to Something Greater

After the *Red Tails* screening, my loyalty felt suspended between worlds, and that feeling was visually and emotionally epitomized by the two primary groups of men who'd attended. One group was entertainment industry talent, whom many would call elites. The kind of people I'd spent the past decade hobnobbing with through my work. Even if they'd suffered hardship, they no longer had to worry about basic needs like accessing adequate health care, paying their mortgage, or being able to afford groceries. They were people living primarily from a place of privilege, who spent their lives flying first-class or on private jets, constantly being interviewed and profiled, basking in public adoration.

The other group was the kind of people I'd become oblivious to: Brave men who'd sacrificed immeasurably for their country, yet had been overlooked and neglected. Men who'd defied stereotypes about Blacks being able to pilot aircraft. Men whose stories went untold, often even in their own families. Men who were only rolled out when someone else could gain something from their presence. Throughout their lives, they were tolerated, but not celebrated. In their final years, they were celebrated, but not compensated.

I now realize that one of the main reasons that elite group existed—was even *able* to exist—is because they were standing on the shoulders of the men in the other group. Just as I was living a life of privilege earned for me by the previous generations' sacrifices. All of which I was completely ignorant of.

I was traveling and partying and laughing all the time, and unable to get a relationship to stick. My life self-admittedly lacked direction and meaning. Many of my friends had settled down, gotten married, and started families, but I liked hanging around celebrities and helping them with their passion projects. As I reflected on my existence, something shifted within me. What was my purpose?

By this point, I'd done well for myself. But I can't say I was doing too well for anyone else. Maybe this was my chance—my shot to make good on the legacy that had been handed down to me.

I was an entrepreneurial thinker and marketer by trade, a professional storyteller. What if I could take that work and turn it into something greater?

This is a question many people ask themselves—right before they start trying to forget about the idea. Nobody wants to see themselves as selfish or falling short, but I most certainly was.

But that one conversation with my dad set me on a different path. If I dedicated myself to the cause of making other people's stories visible, starting with the lost stories of the Davis family, I'd be able to say I had done something right.

The Day Everything Changed

It's not every day you land on page one of the *Wall Street Journal*, but that's what happened to me on February 26, 2012. On that day they ran a story about the red-carpet business I'd founded with my friend William Sidarweck. It was the biggest press we'd ever gotten, and I was excited and proud. I immediately posted the article's corresponding video on Facebook, to celebrate the moment and highlight this validation of the idea we'd spent four years working on.

Before long I received a direct message from James Fenton, the US chief financial officer of the Madison Avenue ad agency TBWA Worldwide. I'd had some dealings with him and the company a few years prior, but I'd never met him in person.

The message said that he'd seen the video on my Facebook page, and he asked me to call him. He wanted to take me to lunch and discuss something. I accepted his invitation, intrigued but with no idea what to expect.

Halfway through that lunch on Fifty-Sixth Street, he asked if I'd be interested in or consider becoming a chief diversity officer.

The serendipity of the moment blew my mind. I'd been so upset about the omissions from the movie. I'd learned a family story I knew next to nothing about. I'd just been alerted to the scale of deletion that had taken place around Black people's stories. I'd realized that staying relevant by going to red carpets all day was feeling superficial and unfulfilling. I'd begun to believe I was wasting precious time in my life. Then I got this somewhat random message from a person

I'd never met in person and had barely worked with, offering me the opportunity to do the very thing I felt called to do.

However, I did have a question:

"What's a 'chief diversity officer'?"

"It's a job where you'd work with us to make sure we're culturally sensitive, that we include people, avoid cultural stereotypes, and are informed. You're going to ensure that the people have a diverse mindset when it comes to our creative product—scripts, castings, commercial scenarios, as well as hiring," he explained.

"Is it full-time?" I still wasn't convinced this was a real thing.

"Yes, it's a full-time job. Every day. This role is part of the executive team."

I left the lunch and talked things over with William. We were best friends and cofounders, and it was important for him to be okay with me considering taking a job and leaving RedCarpets.com, a company we'd built together in the trenches. We created a transition plan, and after we stumbled a bit and resolved some issues, he took over the company and has been chief operating officer ever since.

After the interview process was complete, I signed the paperwork and officially became a TBWA Worldwide employee. I walked into the office on May 1, 2012, my first day on the job, still with no clue what a chief diversity officer was. Nevertheless, I felt certain it was a role that would allow me to advance my new purpose. Over the next decade, I'd figure it out, becoming a leader in the field and building a framework for corporate diversity, equity, and inclusion (DEI).

But that day, I couldn't have imagined what the future held—not only for me but also for the story of Ben Sr. and Ben Jr., the Invisible Generals.

PART I

THE INVISIBLE GENERALS

ONE

Generational Collateral

Our lives are the receipts of our ancestors' journeys—their experiences, investments, relationships, joy, pain, and tragedies.

No one could've anticipated how frigid it would be on March 4, 1873, the day President Ulysses S. Grant was sworn in for his second term. It was the coldest March inauguration on record, and everyone attending was likely bundled up and bracing against the negative windchill. As Grant rode in his custom presidential carriage to the White House, alongside him sat Louis P. H. Davis.

I first learned about Louis Davis from my dad. When I asked other family members about him, I repeatedly heard that his significance could be traced to his purchase of a home in Washington, DC, which would become a launching pad for future generations' success. Ben Jr., Louis's grandson, wrote in his memoir:

> *Louis Patrick Henry Davis, my father's father, had spent his boyhood as a servant in the home of Gen. and Mrs. John A. Logan. General Logan, who later became a U.S. representative*

and a U.S. senator from Illinois, had participated actively in the effort to impeach President Andrew Johnson. My grandfather favorably impressed the Logan family and gradually became their son's companion. Later, as General Logan's protégé, he worked in the Interior Department. He thus gained a measure of economic security and was able to purchase a home at 1830 11th Street NW, where I was born.[1]

Even if you're unfamiliar with Major General Logan, you may know some of the landmarks that bear his name: Logan Airport in Boston, Logan Square in Chicago, and Logan Circle in DC. He was a Civil War hero of the Union Army and became a close friend of Ulysses S. Grant, who had commanded the Union Army. After Logan retired from the military, he entered politics. Between Logan's military prowess, renown, political swagger, and friendship with the president, he was an extremely powerful person and indispensable ally to have—particularly for a Black family in that era.

The record is unclear on when or how Louis became Logan's servant. The term "servant" can mean many different things in a modern context, with both positive and negative connotations. Back then, it meant that the livelihood, accommodations, and tasks were solely determined by the provider. But as a servant, Louis was welcomed into the Logans' home and paid for his labor. He was obviously well thought of and trusted, because he became the companion of Logan's son, John A. Logan II, and was tasked with overseeing Grant's son on occasion as well. So it was noteworthy, but no shock, that Louis accompanied the younger Logan in the carriage during Grant's second inauguration.

Perhaps in appreciation for Louis's years of faithful service, General Logan used his connections to secure a job for Louis in the Department of the Interior. As an assistant messenger for the Office of the

Commissioner of Internal Revenue, he earned a salary of $720 a year (equivalent to about $22,000 today). Once he was gainfully employed, he felt confident enough to start a family of his own. He wed Henrietta Stewart on July 7, 1875, and the newlyweds lived in a six-room brick house near Howard University. In 1878, they moved to another home nearby. Over the years they had three children. On May 28, 1880, the youngest boy, Benjamin Oliver "Ollie" Davis was born.[2]

Louis began making moves into DC's Black middle-class society. In the early 1880s he received a promotion to head messenger in the Office of the Assistant Attorney General, with an annual salary of $840 (around $25,000 today). Henrietta worked as a nurse, and in 1893, the dual income helped the Davises become one of the few Black families in the city to purchase their own home. With General Logan's help, they purchased 1830 11th Street NW—a move that firmly cemented their middle-class status. The Davises owned that home for forty years, and it served as both the launching and landing pad for their family. It allowed them to send their children to school, and it provided financial, and thus familial, stability.

Even though the Davises were progressing, life after the end of Reconstruction proved increasingly difficult for Blacks, especially in the American South. Legal and social pressures, as well as threats of violence, limited Blacks' opportunities. Public accommodations became segregated along racial lines, including in DC. It's important to recognize that the main reason for Louis Davis's exceptional ability to navigate these turbulent waters and accomplish things many Blacks could not imagine during that time—secure a public-service job, acquire real estate, and become upwardly mobile—was because of his powerful ally General Logan.

All these years later, I find it remarkable to contemplate how long my family's influence—over 150 years and across a dozen presidential administrations—can be traced back through America's history.

This knowledge places me in a position of confidence in America, a country where a Black person's citizenship is so often made to feel more like a privilege than a right. We've earned our right to be here as much as any other group.

During Louis's post–Civil War lifetime, America was in a chaotic state that made many Blacks' lives precarious. However, his story highlighted something meaningful to me: not all Black families in the early 1900s were destitute, uneducated, and unemployable. That was the impression I'd gotten from media portrayals of Blacks in post–Civil War America. And Black successes in this era weren't something I was ever taught in school.

Learning his story was evidence that there were beautiful moments where American families of all races could have allies and support, have people looking out for them. People who were similarly pursuing the American dream.

I imagine what it must have been like to consider yourself an American family, yet at every turn, you're reminded that you are Black—a permanent scarlet letter relentlessly bringing to mind that society considers you less than everyone else. And I think about how self-assured you would have to be in that environment, to raise your family with your head up, with dignity, in a country that your family helped build.

Yet Louis wasn't ashamed or afraid to ask for assistance when he needed it, including from people like General Logan. For Louis to get where he wanted to go in life—and to maneuver his children even further down the path—he probably didn't have the luxury of turning down support from powerful allies. The instinct for survival can result in strange bedfellows. Louis likely recognized the inherent unfairness of the world around him, but he also recognized that he had to shape that world to his own needs in order to achieve his goals. While he couldn't reverse or erase centuries of racism and negativity, he could try to make amends in his own way by living an honorable life.

Louis learned the value of the mentor-mentee relationship—the exchange of expertise and enthusiasm for a common good. These rules of engagement are timeless in their effectiveness. Even as a DEI leader today, these principles are the building blocks for good politics. I can be sitting across the table from someone and disagree with them on 99 percent of the issues at hand. Yet if we can put our differences aside and respectfully engage one another, we have a better chance of compromising on the critical 1 percent on which we do agree. That's how the needle gets incrementally moved.

In my mind, I often reenact how incredible it must have felt when Louis rode to the White House in that carriage with President Grant. I feel that this moment—one that I can only imagine was filled with excitement, serenity, or even anxiety—must have been a turning point in his life, when his energy and the exuberance of youth propelled him forward. He'd done it. He'd make a name for himself and for his future family. He'd seize every opportunity and access to every resource. He'd buy a home to provide safety and security for his children and send them to the best school district for Black kids. I love him for all of that, and not just because my life is still benefiting from it today. Louis is an inspiring example of perseverance. His ability to look ahead and stay the course into an unknown world for him and his family, with no safety net, is a seed planted in the soil around the base of my family tree.

Louis laid the foundation of generational collateral for my family. I define that term as the assets, information, knowledge, stories, experiences, DNA, health, mental and physical wellness—or, on the flip side, the pain, suffering, despair, and poverty—that is passed down to us from previous generations. The concept has become near and dear to me and gives me much more empathy and understanding toward my family's journey.

Through his example, Louis demonstrated for others, foremost his children, what it meant to work within a system to help evolve it—to build political and financial capital. One thing I learned the most from Louis was that simply being in the room was powerful, and even though he didn't choose to be born a servant, he was shrewd and ambitious enough to leverage his proximity to those who had both power and influence, so his family could succeed in ways that otherwise might not have been possible. He refused to back down to society—society would have to catch up with him.

Louis's youngest son, Benjamin—affectionately known as Ollie—took these lessons to heart. Louis hoped his young son would follow in his footsteps and work for the government. Henrietta wanted her son to become a minister. Ollie had other ideas. He wanted to enlist in the Army. His desire to be a soldier blossomed at M Street High School, where he was first a cadet and then a member of a Black unit of DC's National Guard.[3]

Ollie's parents were disappointed, but saw the passion and virtue in his decision. Louis set out to do what he could to make it a reality, and he reached out to General Logan, to see if an appointment to the United States Military Academy at West Point could be secured for Ollie. West Point attendees were commissioned upon graduation, which meant they automatically attained the rank of second lieutenant—unlike enlisted men, who had to work their way up through the ranks.

Louis received word that, for political reasons, President William McKinley wouldn't appoint a Black man to West Point.[4] While this was devastating, Ollie wasn't deterred and soon after opted to join the Army as an enlisted man in 1898, a plan Louis stringently disagreed with, feeling that being a private, instead of an officer, was beneath his family and his son. The pay was minimal, and the military was rife with racism. Black soldiers were relegated to "pick-and-shovel jobs," a term to describe roles that didn't allow for advancement through the ranks. Every day Ollie would be faced with what my family would call "indignities," which included disrespectful attitudes, tones, and behaviors in response to his race. Indignities were something my family—including me—was taught to understand, live with, and move past with professionalism.

Ollie quickly made strides in his career. One reason was that he could read and write, unlike many of his fellow soldiers.[5] As Ben shared, "His knowledge of Army regulations and administration and his willingness to work long hours made him an indispensable asset to his organization and enabled him to advance to sergeant-major in only a few months."[6] Even if Louis and Henrietta didn't want their son to become a soldier, through their military and political connections and familiarity, supportive parenting, and commitment to educating their children, they had positioned him to excel on any path he chose.

Last year, I discovered that I could go online and access many items in Ollie's archives, which are cataloged at the US Army Heritage and Education Center, at the Army War College in Carlisle, Pennsylvania. One of the most fascinating and unexpected items I discovered was Louis's will. Since Louis died before my dad was born, the only information I know about him is secondhand; I was excited to see a document that he himself had touched.

He left everything to his wife, Henrietta, and indicated the circumstances under which his children would inherit his estate, which included money, rail stocks, and the DC home. After his death, everything was to be divided among the children. But to me, the most interesting part about Ollie was as follows: "To my son B. O. Davis I bequeath five (5) dollars, to be paid out of my estate, he having received his full share during my life time in money and in my efforts to secure him his position in the U.S. Army."[7] Louis might have opposed his son's chosen path, but he'd obviously played a significant role in Ollie's launch from the landing pad of the Davis home to a successful military career.

Louis Davis, Ollie, and Ben all called 1830 11th Street NW home at some point in their lives. I once tried to visit the home, so I could see this property that has been so influential in my family story. I located the site and saw it had been rezoned for condos. There was no trace of my family home. It's too bad the property wasn't passed down to my dad, but the resources it provided changed my family in a way that transcends material gain. The foundation the home provided to allow future generations to flourish in a country that wasn't sure—and sometimes still isn't sure—how to recognize Black achievement cannot be overstated.

TWO

Extraordinary Just to Be Ordinary

Previous generations lived extraordinary lives
so we can live ordinary ones.

As an enlisted man, Ollie's reputation for being a hard worker preceded him, and his transformation into a trailblazing officer began right out of the gate. Within his first year, he found himself out west in the 9th Cavalry, in one of four all-Black regiments in the United States. Part of their mission was to protect US interests during westward expansion, which was code for fighting Native American tribes. The 9th became the famous "Buffalo Soldiers" because the Native American tribes they fought thought the Black men's dark, curly hair was reminiscent of a buffalo's coat.[1]

Looking back at this moment, I imagine the uncompromising position they must have been put in, being only a handful of Black soldiers in the Army and assigned the task of pushing back the Native Americans.

Their equestrian skills were so exceptional that, in 1907, some of the Buffalo Soldiers were sent to West Point to deliver cavalry training, a practice that would continue for forty years thereafter. Because of their race, many of the Buffalo Soldiers' deeds and accomplishments went unknown and unrecognized for decades. Thankfully, that's finally begun to change. In 1973, the campus location where they conducted training was renamed Buffalo Soldier Field, and in 2021, they were honored with a statue at West Point.[2]

Ollie's fascination with horses and cavalrymen began in his school days. In addition to reading books like *Black Beauty*, his interest was piqued by seeing the 9th Cavalry soldiers when the unit participated in the 1893 parade for President Grover Cleveland's inauguration in Washington, DC, which Ollie attended as a spectator.[3] His passion for horses and all things equestrian would become his calling card, and he knew that if he became the best cavalryman in the history of the US Army, it would be difficult for the powers that be to deny him the opportunity for promotion.

While serving out west, First Lieutenant Charles Young, West Point's third Black graduate, became Ollie's mentor, and Young's profound impact undergirded the young soldier's accomplishments. Young was a positive figure in Ollie's life and encouraged him to excel. He also shared his experience as an officer, and tutored Ollie in math, to help him ace the commissioned officer test. One of the most significant teachings Young imparted to Ollie was that excellence in executing one's tasks was only part of the success formula. You had to equally navigate the political part, to outsmart and outmaneuver those who tried to obstruct you and prepare for those you didn't see coming. Along with that mental acuity, you needed emotional fortitude. Additionally, Young's insider knowledge of surviving West Point as a Black cadet would be passed down the Davis line to Ben Jr. Ollie was adamant that this moment in time was a sign of change in America, so he absorbed and internalized the way of thinking that Young shared.

Ollie ranked up to sergeant major in a matter of months. Then, in 1901, only a couple of years after he'd enlisted, he was recommended for a Regular Army commission as a second lieutenant of cavalry. One of his commanding officers, post commander M. B. Hughes, said the following in his recommendation letter: "He possesses the honesty, sobriety, integrity, and intelligence which is so necessary in a commissioned officer, and if commissioned in the army will fill the position to the satisfaction of the country, and with honor to himself."[4]

You have to appreciate how incredible this was. At the time, the only other commissioned Black officer in the United States was First Lieutenant Charles Young. Since Young had graduated from West Point, he emerged as a commissioned officer upon graduation; he didn't come up through the enlisted ranks the way Ollie did. President McKinley, who had denied Ollie entrance to West Point, ended up being the one who signed him into officer status a mere two years later. It's something President Bill Clinton alluded to in his speech at Ben Jr.'s four-star-general ceremony in 1998: "[Ollie] enlisted in the army and distinguished himself immediately. In less than two years he was an officer. It takes longer if you go to West Point."[5]

His son, Ben Jr., had been immersed in all things military his entire life, yet even he didn't know how his father had accomplished this extraordinary feat. Ben Jr. wrote in his autobiography, "My conclusion, very simply, is that he was a most exceptional young man, capable of convincing many in the racially unenlightened Army chain of command that he had earned and deserved a Regular Army commission."[6] Ben Jr. loved telling the story of his father being commissioned, his voice infused with pride.

In exploring Ollie's archives, I found a letter he'd written to his father, Louis, which I think provides some insight into his seemingly inexplicable ascent through the ranks. With the letter to Louis, Ollie enclosed two letters of recommendation, the one from Commander Hughes as well as one from First Lieutenant Young, and wrote:

> *I thought I would send them to you as it may enable you to get some of your friends to work for me in Washington. . . . Use all the pressure that you can. If I win this means a fortune for life and if I lose, it will be no disgrace. . . . Hoping that you will help me all you can and that you realize the fact that should I be fortunate enough to gain the position, it will not only be as fortune and honor for me, but the whole family.*[7]

Ollie understood what was at stake in this game, and he recognized that as much as his disgrace was the family's disgrace (hence using the name Benjamin when he enlisted instead of Ollie), his honor was the family's honor. And he obviously had a level of awareness that his father's reputation and connections were valuable.

All these behind-the-scenes machinations and calling in of favors may be what Louis referred to in his will, when he referred to his "efforts to secure him his position in the U.S. Army." Though Ollie greatly benefited from the generational collateral of a stable home and family, he may have benefited even more from the generational collateral of proximity and connections to people with power and influence.

While Ollie made great strides in his military career, he endured hardship in his personal life. I think the personal element of these historical stories is often overlooked, as well as the complexity of people's lives, which is unfortunate, because the personal humanizes these stories. When we idealize our historical heroes instead of showing them in full relief, they're less relatable, and it becomes harder to see ourselves in their lives and struggles—and their victories. No one's perfect, and lauding someone's achievements doesn't mean you completely overlook their flaws and personal struggles—the objective is taking what we can from those who came before us, to improve our lives and the lives of those who come after us.

Ollie's wife, best friend, and partner of ten-plus years, Elnora, died nine days after giving birth to their third child, a daughter also named Elnora. Suddenly, he was a widower, a single dad with three young children to care for. He was on duty as a professor at Wilberforce University in Alabama, but he could be called back into action at any time. Sure enough, less than a year after Elnora died, he received orders to the Philippines—ironically assigned to the USS *Logan*—for what would end up being a three-year tour. Some of the children's aunts offered to take one child each, but Ollie felt strongly about keeping them together. The three children, including Ben Jr., stayed with Louis and Henrietta in DC.

In the midst of Ollie's grief over losing his beloved wife, he faced a long-term separation from his children. Thank goodness he could rely on his parents to house and care for his kids, which undoubtedly gave him a measure of peace. As committed as Ollie was to being an excellent soldier, he was equally committed to being an excellent father and doing the right thing for his children, and the narrative of what he stood for. A strong, confident, extraordinary man, who's also a single dad fighting every day to keep his children together and safe, wasn't common in the press or among those who stereotyped Black Americans.

Sometimes, being extraordinary isn't only about professional accomplishments. It's about the sacrifices in our personal lives and relationships as well.

While Ollie was stationed in the Philippines, he kept up a lively correspondence with Sadie Overton, a young teacher from DC's middle class, who had helped with the children after Elnora's death. He planned to marry her after he returned to the States, but the Army delayed his return. Therefore, he sent for her, and they got married in the Philippines in 1919.

When they returned to the States in 1920, Ollie was assigned to teach at the Tuskegee Institute, to serve as a professor of military science and tactics. The family was overjoyed to be reunited, and the children were thrilled that Ollie had taken Ben's suggestion that he marry the woman they called "Mother Sadie."

In spite of the family's happy reunion, the dark cloud of racism often hovered over their life in the segregated Deep South of Alabama. Early in Ollie's tenure, the faculty and staff learned that, close by the institute, a new Veterans Administration hospital was opening that would cater to Black veterans. In response to the news that the facility's doctors and nurses would be Black, the Ku Klux Klan planned to march past the institute, which had instructed its employees to stay indoors with the lights off, to avoid provoking the Klansmen. Ollie was having none of that, and Ben recalled the following about the incident:

> *But as a Regular Army officer, my father refused to cower behind closed doors. Furthermore, he believed that the entire Davis family should make known its opinions of the Klan by staying visible and not hiding in the shadows. On the night of the Klan, therefore, Mother Sadie, my two sisters, and I sat quietly on our porch, my father resplendent in his white dress uniform, and the rest of us viewing the parade with some concern about what might happen when the Klansmen marched by. Our porch light was the only light to be seen for miles around except for the flaming torches of the Klansmen, who passed by on the street only a few feet from where we sat. Also*

resplendent in white—robes, masks, and hoods—they passed by without incident.[8]

Among the porch light, the flaming torches, and the white outfits of the Klansmen, nothing shone brighter that night than a Black man in a pristine white military uniform. His courage in the face of racism, and his refusal to be intimidated modeled for his children what it means to be a disruptor. It's a lesson that undoubtedly came to Ben Jr.'s mind often during his four years at West Point. But in spite of the persecution, he was determined to fulfill the promise he'd made to his father to become America's greatest general.

False narratives about Blacks emerged from many sources—including the federal government. In November 1925, the US Army War College issued a damning report on the state of Black soldiers, stating that they were inherently inferior. Not one doctor helped create this report, but it was considered an authoritative source, cementing a narrative that society fed off. It was confirmation bias at its worst.

I've become quite interested in the document over the years—so much so that I even traveled to Carlisle, Pennsylvania, to visit the Army War College. I needed to see the condemning words on the page with my own eyes because I found it impossible to wrap my mind around the reality that this report existed. The moment felt surreal, sad, unbelievable, disturbing, and ironic, considering that this document is housed in the same location as Ollie's memorabilia.

Among its many lies, the document asserted the following:

In the process of evolution the American negro has not progressed as far as the other sub-species of the human family. As a race he has not developed leadership qualities. His mental inferiority and the inherent weakness of character are factors that must be considered with great care in the preparation of any plan for his employment in war.[9]

With these kinds of findings being touted as accurate, is it any wonder that both Ollie and Ben Jr. met such scrutiny and resistance throughout their military careers?

For obvious reasons, the War College report was kept highly confidential. Only a select handful of individuals knew of its existence, and those in the know leveraged it to fight integration of the armed forces. Because military policies typically set the tone in different sectors, such as education and business, the "justified" segregation had a ripple effect into these other arenas, limiting opportunities and snuffing out the dreams of many Blacks. This one document damaged public perception of Blacks for decades.

Ben Jr. didn't hear about the report until years after it was put forth, and he wrote the following about it:

It concluded that the intelligence of black people was lower than that of whites, that blacks lacked courage, that they were superstitious, and that they were dominated by moral and character weaknesses. It also stated that the "social inequality" of blacks made the close association of whites and blacks in military organizations "inimicable to harmony and efficiency." The Army had approved this "study" and used it as the basis of its discrimination against blacks.[10]

It's unlikely Ollie knew of the report. Even if he didn't, he experienced the discriminatory aftereffects. Though he was able to climb up through the ranks, every step of the way, he had to fight not only for the higher titles he deserved but also for the commensurate duties. At each rung on the ladder, he'd encounter white soldiers who were uncomfortable with—or outright antagonistic about—reporting to a Black officer. The War Department (now known as the Defense Department) didn't want him in a position where he'd outrank a white officer. This placed a cap on Ollie's horizons and limited his ability to pursue his preferred service in the field with troops, alongside his fellow soldiers.

Fueled by his tenacity, discipline, and commitment to being the greatest—and with a touch of desire to silence his detractors—Ollie persisted in the military. His patience paid off when President Franklin D. Roosevelt promoted him to brigadier general in 1940. He was the first Black officer to attain this status in the United States. Among other accolades, he received a Bronze Star Medal and the Distinguished Service Medal. His citation for the latter stated, "The initiative, intelligence and sympathetic understanding displayed by him in conducting countless investigations concerning individual soldiers, troop units and other components of the War Department brought about a fair and equitable solution to many important problems which have since become the basis of far-reaching War Department policy."[11]

Like his father, Louis, before him, Ollie was determined to pave the way for the next generation, first and foremost his son. Ollie would continue to build on the generational collateral he'd received, to improve his son's opportunities and chances for success. When he became an officer in 1901, he had no way of knowing there wouldn't be another Black officer commissioned until 1936. That officer was his only son, Ben Jr., who would navigate many of the same obstacles his father had on the way to achieving the status of general.

THREE

West Point's Invisible Alum

The stone the builders rejected would one day become the cornerstone.

Similar to the way Ollie had developed a passion for horses and cavalry as a youth, Ben Jr. fell in love with airplanes and aviation. In 1926, when Ben was about thirteen years old, Ollie paid five dollars for the pilot at a barnstormer show at Bolling Field, outside Washington, DC, to take Ben up for a ride. Though five dollars doesn't sound like much (it's equivalent to about eighty-five dollars today), back then that was a lot of money for Ollie to invest in something. But that decision would alter the trajectory of his son's life, as well as that of my family and the country. In his autobiography, Ben said of his father's decision: "I can only guess that he was looking far into the future and, seeing airplanes in that future, realized in some mysterious way that I would benefit from the experience."[1] From that moment on, Ben knew his life would never be the same: "About all I remember are the takeoff and the feeling of exhilaration at being in the air, looking down on the city of Washington and up at white clouds far above us. And I remember a sudden surge of determination to become an aviator."[2]

Aviation was a relatively new field, and the private sector was mostly segregated. Thus, his only hope of becoming a pilot was as a military officer. Attending West Point was the most surefire way to become an officer. It was a means to an end.

Unknown to Ben, Ollie had begun behind-the-scenes machinations to ensure that his son would eventually be placed at West Point. Ollie wrote a letter to the lone Black member of Congress, Oscar De Priest, a Republican from Illinois. He expressed his concern that after he retired, the Army wouldn't have a single active-duty regular officer, and he suggested Ben as a possible West Point appointee.

There was one catch: a congressperson could only appoint someone to West Point if that person lived in their district. Upon learning this, Ollie and Sadie made the difficult decision to send Ben to live in Chicago in 1930. He rented a room from a family while attending classes at the University of Chicago. A year later, in 1931, when he sat for the three-day examinations required for West Point, he failed the English history and European history portions of the test because he hadn't studied these subjects.

He dreaded telling his father what had happened: "Failing the examination was worse than an embarrassment; it was mortification. I apologized for letting the family down and confessed that I was not the man I had considered myself to be. I asked that he renew his faith in me and give me another chance. He must have been as thoroughly embarrassed and disappointed as I was, but his reply was a classic letter of support, indicating his continued faith in me."[3]

Receiving official notice of his failure made Ben even more determined to pass the examination and enter West Point. He wrote, "It was on this day that I fully resolved to go to West Point, graduate, and seek a career in the Army Air Corps. I never had further doubts of any kind about my future, which I was completely convinced lay in the military service."[4] He doubled down on preparing for the examination and dropped out of the University of Chicago to fully focus on the effort. The second time, he passed with no problem.

After Ben was officially appointed to West Point, he received another supportive letter from Ollie that included a reminder: "Remember twelve million people [the Black population of the United States] will be pulling for you with all we have."[5] And Ben was ready to make his father proud: "In climbing through the Army's ranks from 1898 to 1932, my father had overcome what seemed almost impossible odds. In spite of the attitudes of whites in the United States toward all people of color, he had managed to buck the system and accomplish his goals. He had made life easier for me. Now it was my turn to make things better for those who would come after me."[6]

Ben eagerly anticipated his arrival at West Point in 1932, excited at the prospect of making lifelong friends, sharing experiences during rigorous training, and becoming an officer with his class. When he arrived on campus from the train station, he quickly realized things would be different. For starters, he was called into the commandant's office and told he was being assigned a "special room." Come to find out, the white cadets refused to room with him, and his solo bunk would completely isolate him.

Since the War College report was issued several years prior to Ben's enrollment at West Point, one can surmise that any administrator who read it might have used the information to excuse the way he was treated during his time there. After all, why would they want to support and encourage a cadet who would never amount to anything, who was inferior to his white counterparts and would never be capable of leading?

Additionally, Ben discovered he'd be on the receiving end of a cruel West Point tradition called "silencing," where a cadet was only spoken to in the line of duty. In practice, silencing was psychological, mental, and emotional torture. Unlike the other cadets, he had no roommate for companionship, and in four long years no one welcomed him at their table in the mess hall, so he ate his meals alone.[7] Some upperclassmen seemed to think Ben needed to be "put in his place"—this son of a military man who may have been perceived to be riding on the accomplishments of his surname. That kind of patronage, imagined or not, couldn't be tolerated for a Black man. They sought to make him so miserable that he'd leave West Point. Or, as they said in the military, they wanted him to "wash out," the same fate previous Black cadets had met. But they'd severely underestimated their target. How the story is retold and how I imagine it are two different things. No one talked to him for four years? How was this possible?

"What they did not realize was that I was stubborn enough to put up with their treatment to reach the goal I had come to attain," he'd write. "I made my mind up that I would continue to hold my head high. At no time did I consciously show that I was hurt; even at this early date, I took solace in the fact that I was mature enough to live through anything other people might submit me to, particularly people I considered to be misguided."[8]

Ben also made the conscious choice to focus on the benefits gained from this treatment. He said that if they hadn't silenced him and acted like he was invisible, they would've tormented him instead, and that probably would've resulted in him washing out. In this case, being invisible was an opportunity—one that allowed him to quietly break down barriers then and in the future. Since he was left to his own devices, that allowed him the opportunity to study and read everything he could get his hands on. The only courses he performed poorly in were those impacted by the silencing—the ones that required his classmates to engage with him. He also had ample time to maintain a lively correspondence with his father and Mother Sadie, and hearing news of their adventures helped him feel less isolated. He withheld his difficulties from his parents, because he didn't think revealing them would do any good. Having a father who was a military officer likely offered him a certain level of protection too.

One bright spot during this season in his life was his romance with Agatha Scott, a young woman he'd met at a dance in New York City while on Christmas leave from West Point. Their courtship began in 1934, and as their relationship progressed, Agatha would drive to West Point almost every weekend to visit him, providing the conversation and companionship he desperately lacked and needed. Ben fell in love with Agatha and wanted to marry her. However, cadets were prohibited from getting married, so Ben had to wait until after graduation to make Agatha his wife.

Ben rarely spoke of his West Point experiences with the rest of his family, including me. He seemed to feel that mentioning the ordeal in any detail would be like reliving it, and his generally positive and optimistic attitude wouldn't allow for intrusive memories of such abuse and the subsequent trauma. His behavior was the same as the previous Black West Point grad, Charles Young, who'd likewise endured and survived the torment inflicted upon him. In a eulogy W. E. B. Du Bois wrote about Young, he said, "No one ever knew the truth about the Hell he went through at West Point. He seldom even mentioned it. The pain was too great."[9]

Most of what I know about Ben's experiences at West Point is based on what I read in his autobiography and what Agatha shared with my father. Ben did tell us how it was hard for him to understand how his fellow cadets, who ostensibly believed in West Point's motto—duty, honor, country—the same as he did, could treat a fellow cadet that way.

He said the mess hall "situation" was the worst embarrassment of his life. First-year cadets, called "plebes," had to ask permission to sit at someone's table to eat. Of course, everyone refused to let him sit with them. Even as an upperclassman, if he sat at a table, the other cadets would stand up and leave, so he'd have to eat alone. He endured this indignity three times a day, fifty weeks a year, for four years straight.

Another situation he spoke of was attending the annual Army-Navy football game. All the other cadets took the same buses on the hour-plus ride to the stadium, but Ben rode in a separate bus, with only the Black bus driver for company. By his second year at West Point, word got out in the Black community about the lone Black cadet. People, both Black and white, would stand outside the bus and ask him for his autograph, which continued through his third and fourth years. Though his classmates ostracized him, the Black community continued to embrace him and honor his struggle. Ollie's assertion that America's twelve million Blacks were rooting for Ben had proven true.

Amid his suffering, his resolve and integrity shone through, and he offered a clue to its source when he wrote, "I never felt sorry for myself. I knew I could push aside any obstacle in my path. My father had taught me to be strong; he had endured adversity, and so could I."[10]

After I started engaging with West Point in 2015, a refrain I heard from many people across levels and ranks was that no one could grasp how Ben was able not only to survive in that environment but also to thrive, graduating thirty-fifth in his class of 276. His ranking likely would have been higher, if not for those course grades that suffered as a by-product of the silencing. When they hear Ben's story, cadets, faculty, and graduates alike tell me that he must have been the strongest person who'd ever graduated from the Point.

Even Ben's peers seemed to have been won over by his unequaled ability to transcend the horrific treatment. In the 1936 edition of *The Howitzer*, West Point's yearbook, the following was written about Ben: "The courage, tenacity, and intelligence with which he conquered a problem incomparably more difficult than plebe year won for him the sincere admiration of his classmates, and his single-minded determination to continue in his chosen career cannot fail to inspire respect wherever fortune may lead him."[11] The dissonance between his daily reality and this glowing statement blows my mind.

I wish I'd talked with Ben more about his time at West Point, even if he was unwilling to share much. In my youthful ignorance, I didn't have an appropriate level of empathy for him and what he'd experienced, whether at West Point or throughout his military career. He'd always been kind and generous to me and was one of the most joyful people I knew. It was hard for me to fathom how someone so nice could have been treated so horribly all the time. He wasn't combative and didn't retaliate. He was merely a young man trying to chase his version of the American dream, just like all the other cadets.

In his typically pragmatic way, Ben sometimes told me that as awful as life could be as a Black soldier, it was preferable to life as a Black American in the private sector. In the military, at least you got paid a guaranteed equal wage to whites, had a roof over your head, ate three meals a day, garnered marketable skills, and had a purpose and mission in life. And if he wanted to pursue his desire to become a pilot, the only place he could do that was in the US military, where he'd also have access to the world's cutting-edge aviation technology and aircraft. Conversely, the segregated private sector opted not to train or hire Black pilots. Nor did they intend to.

The year Ben graduated, General John J. Pershing was serving as West Point's superintendent. Pershing had commanded the US Allied forces during World War I, and his history with Black troops was complex, to say the least. In a private correspondence about Black troops that had been sent to the French military, dated August 7, 1918, Pershing wrote the following: "We must not eat with them, must not shake hands with them, seek to talk to them or to meet with them outside the requirements of military service. We must not commend too highly these troops, especially in front of white Americans."[12] Since Ollie had served in the military throughout World War I, he knew of Pershing's reputation, and there was speculation that the superintendent might not shake Ben's hand when he presented him with his degree and commission.

In a historical moment, Pershing did shake Ben's hand—forty-seven years after Charles Young had graduated from West Point—and Ben received thunderous applause from those in attendance. Newspapers across the country featured his graduation from the United States Military Academy, and a Black magazine, *The Crisis*, placed one of Ben's cadet photos on its cover, proclaiming him the "No. 1 Graduate of the Nation."[13] Of that day, Ben wrote, "I felt that my graduation represented a watershed in my career and that the worst was over."[14]

Two weeks later, on June 20, 1936, Ben and Agatha married in the West Point Cadet Chapel. Although it was tradition for the cadets to attend one another's weddings, Ben continued to be shunned after graduation, when not a single classmate attended his wedding. The only attendees were the chaplain and three family members: Ben's sister Elnora, Agatha's mother, and Agatha's sister Mildred. From the outset, it was clear that Ben and Agatha's life would often be them against the world, but they knew they could weather any storm together.

America now had two Black military officers, a father and his son. This was a major accomplishment for both Ollie and Ben and for my family, and a huge step forward in Ollie's dream to raise the nation's greatest military general. The foundation for the family's—and the country's—further evolution was set.

Seeing the picture of Ollie and Ben shaking hands at graduation has been one of my greatest moments of discovery: the image of a father and son celebrating such an amazing accomplishment, with only the son fully knowing what he'd endured to make it to that point. I've stared at this photo for hours, imagining what that day was like and all the things that were both said and unsaid.

FOUR

Living Life in a Liminal Space

Playing the long game means being prepared for the good, the bad, and worst of all—the indifferent.

Between the two world wars, the US military had created a problem of its own doing: the two highest-ranked Blacks, Ollie and Ben, couldn't command white troops in a fully segregated military. The US War Department had to do something with them, so the two Black officers were given what was considered a throwaway assignment: visit Black colleges throughout the South, to teach military science and work with the ROTC (Reserve Officers' Training Corps) students, to train the next generation of Black soldiers. From 1937 to 1940 they'd travel together, as their regular duties permitted.

During a couple of these years, Ben was serving on the faculty at the Tuskegee Institute. Both Ollie and Ben had a long-standing and deep connection to the Tuskegee Institute. Ollie had two stints there, running the ROTC program. His first tour of duty was from 1920 to 1924. Ben was sent to Tuskegee in 1938, to replace the person who'd taken over after his father's second round there. It seemed that history was repeating itself, as both Davis men were moved into teaching positions because the military didn't know what else to do with them.

While the institute's campus was a flourishing hub of Black education and excellence, the town beyond remained hostile. As Ben described it, "Tuskegee Institute was an island within which blacks could live and move about comfortably, surrounded by whites who handicapped themselves and poisoned their own individual and collective lives by the virulence of their hatred of blacks."[1] Returning there as an adult, the childhood memory of that Ku Klux Klan march past their home undoubtedly came to mind on occasion.

Both Ollie and Ben were experienced teachers and had much to offer in the way of information. But what their students received was above and beyond the required instruction: they received inspiration and encouragement from two men they considered heroes, Ollie and Ben. And the students needed this interaction to understand what had elevated these two strong, smart men. Ollie and Ben told them to be positive and not complain. To stand up for themselves, to sit up straight and look people in the eye. Not to use all their energy on rhetoric and talking and being a scholar. To be above reproach and set an example of undeniable excellence through performance. They enchanted the students with the power and magic of possibility, something all young people need—particularly Black men living at a time when rampant racism was alive and well throughout America.

Two realities coexisted each and every day during these tours, and Ollie and Ben existed in this liminal space. The military establishment wanted nothing to do with them and shuffled them from one disappointing post to the next. Yes, they were as accomplished as any Blacks in the military had ever been, but no white soldier could salute them, and they couldn't command white troops. Yet they chose to tell these students the sky was the limit, knowing full well the many obstacles and limitations Blacks faced in the military. Ollie and Ben were training and inspiring the young men for an opportunity that wasn't around at that time, but might manifest in the future.

Ollie and Ben knew in their minds, and through their intel, that it was only a matter of time before the next conflict erupted, and the United States would likely be compelled to join in and defend itself and its allies. For that, they rigorously trained these young men not only for combat in war but also as leaders for roles that may or may not become reality. The process they undertook was like watering seeds that would later blossom. You never know which seeds will flourish, so you treat them all the same, with equal water, fertilizer, and sunshine. You operate as though every single seed will grow and flourish.

Over the years, Ollie and Ben built those relationships with the students, fostering trust and camaraderie. They told these men that their nation would need them one day. And they must be ready for that moment.

Toward the end of these college tours, something happened that they couldn't have planned. On October 25, 1940, my dad's seventh birthday and twelve days before the big election, President Franklin D. Roosevelt promoted Ollie to brigadier general. It was a moment Ben fondly remembered, as he recalled: "Although the promotion was motivated primarily by the hope of winning black votes in the 1940 presidential election, my father had richly deserved it for many years."[2] Ollie received his orders to report to Fort Riley, Kansas, as a commander of two Black cavalry regiments. One of his first acts was to request that Ben be reassigned from Tuskegee Institute, to serve as his aide.

Although both Ollie and Ben desired more active roles in the military, they undertook every role with pride and served with excellence, regardless of what they had to walk through, personally or professionally.

They understood that complaining about their circumstances without the animation of personal agency would neither convince others they were their equals nor result in change. In many ways, they might have been denied a voice, but they could "speak" through their mere presence. As men of color, they recognized that being in the room was disruptive in and of itself. Their visibility put the entrenched powers who currently dominated the table on notice, and lent hope to those who dreamed of joining that table in the future—colleagues who until then hadn't seen someone like Ollie or Ben in a position of such great power.

Nevertheless, every voice is not a vote. Time and again, Ollie and Ben saw powerful forces invite voices to the table without extending the privilege of decision-making. Those in power tend to fiercely guard their own status, which was one of many reasons Ollie and Ben emphasized performance. It became your voice when one was denied to you, and it gave you a way to break down barriers when words proved inadequate. Ben in particular adhered to this path. In his mind, his performance was both the "voice" and the "vote" that would change the world for the better.

Even though the military higher-ups likely didn't expect anything good to come from Ollie and Ben's unofficial recruitment assignment, it had positive outcomes. To me, the most important one is that it further reinforced Ben's purpose. While his ultimate goal was to become one of the greatest US military generals, an intermediate and related purpose was to inspire a generation of Blacks through performance—to lead by example and show the entire country that Black soldiers were just as capable as their white peers.

Getting a PhD in Black America

As I'd come to learn, Ollie and Ben both attended Black colleges—Ollie took courses at Howard University, and Ben spent a summer at Fisk University in Nashville, Tennessee. They were familiar with higher education and were provided the skills they equipped their students with. Not to mention that Ollie had taught at Wilberforce University, and of course, both Ollie and Ben had taught at the Tuskegee Institute, which later became Tuskegee University. Today these schools are designated and known as HBCUs: Historically Black Colleges and Universities. Although my dad went to New York University, he also attended Howard University School of Law in Washington, DC. My mom went to Stanford University and NYU, but also attended Hampton University in Virginia.

My whole life, my parents had encouraged me to be aware and proud of my culture, but they didn't want me to attend an HBCU, or a school in Florida, Hawaii, or Connecticut. They clearly had their prejudices, though I didn't know that this was because of their own educational views on opportunity. I didn't give my parents' resistance much thought, because I was learning from them without being combative, but when I started the process of reclaiming my family legacy, I started researching the history of HBCUs and wanted to better understand their role in my family story. That exploration led me to discover my family's long-standing and deep connections to the Tuskegee Institute, as well as more information about Ollie and Ben's unofficial recruitment efforts. Ollie and Ben's exclusive focus on the HBCUs resulted in one of the biggest, most critical opportunities for the Black community to demonstrate the skills they already possessed—and soon they'd have a chance to show the entire world their excellence.

The summer between my junior and senior years at Syracuse University, I had the chance to intern for music mogul Quincy Jones, at

Qwest Records in Burbank, California. Even though I was a lowly intern, he treated me with respect and spoke to me like I was one of his students or his own child. He'd talk in sound bites of inspiration, knowledge, and wisdom. One day he broke down the phenomenon of one-hit wonders, which resulted from a combination of talent, luck, and timing. He explained that a one-hit wonder was a song whose aura was bigger and more powerful than the artist themself. Most musicians who have a one-hit wonder say it's both a blessing and a curse. Some make a career out of it, while others get engulfed in it and feel trapped. Most musicians would give anything for just one of their songs to be a hit. Yet when a song takes on a life of its own, it can eclipse anything else the artist accomplishes in their whole lifetime. The story of the song becomes the artist's story, inextricably linked to them forever.

Recognizing the power of controlling the narrative was a motivator for me to enter the media industry. After the *Red Tails* screening, I began diving into the disparity in media portrayals of Black Americans throughout history. I couldn't undo the wrongs of the past, but I had an opportunity to create a narrative that could serve as an antidote to false and negative perceptions of Blacks.

The foundation for this work was laid after I graduated from college. Two important mentors, Tommy Hilfiger and Earvin "Magic" Johnson Jr., helped me better understand the worlds of sports, business, fashion, and music in American culture. These life experiences taught me the importance of being humble and relatable, and also equipped me to be an advocate in my work as a diversity officer.

I initially got connected to the Hilfiger family during the internship I had with Quincy Jones. Tommy and Andy Hilfiger had created a partnership with Quincy on a record label. In the late 1990s and early 2000s, the Hilfiger brand was ubiquitous. They were creating content and shaping culture through style, sound, and personalities and could work any medium. You couldn't turn on a TV or read a magazine without seeing the brand.

When I began working with them, both brand namesake Tommy Hilfiger and Andy Hilfiger, his brother and business partner, took me under their wings. Well beyond my time with their company, these two brothers—and even their sisters, Ginny and Betsy—became my mentors across many areas of my life. My mentorship with them taught me so much about family, relationships, pop culture, Americana, entrepreneurship, adulting, and how to move properly through rooms. We had conversations about building a brand and how to navigate different elements of business. This was perhaps most evident in their work with Tommy Jeans and the way they built an inclusive style that captured the heart of a generation. Tommy's most essential lesson for me was to live your life like a CEO. He told me not to start at the bottom and work my way up—start at the top and find a way in. Then he explained: When you go to school, they teach you how to be an employee—how to work for people, take direction, etc. If you want to be a game changer, you have to think like a CEO. You must be able to look down the street and around the corner. To think about your fee, your product offering, and your specialty. Tommy put his money where his mouth is when he gave me funds to start my own business.

I soon used these funds to launch my own company, which offered marketing and advertising services geared toward growing communities. From there I was recruited to be the president of one of Magic Johnson's companies, a role I consider the centerpiece of my experience.

Similar to the potential Ollie and Ben saw in those college recruits, Magic saw some of the same potential in me. Though I wasn't as undeveloped as a seed, I definitely needed some serious fertilizer and pruning. Left to my own devices, I would've been showing up to meetings in scuffed sneakers with a backpack slung over my shoulder, like some recent college grad. The fact that he believed in me, and that his faith in me never wavered in spite of my missteps, was a big deal. To this day I have immense appreciation and gratitude for him.

Magic recognized that Black America was an underserved and domestic emerging market, with an opportunity to leverage capital and generate wealth that could be continually reinvested in the community. This market was frequently overlooked by mainstream businesses. Working with Magic, I had the then-rare opportunity to see the advantages and reach of a responsive, Black-owned business. His faith in the Black community wasn't entirely dissimilar to the potential Ollie and Ben saw in their recruits, who were overlooked by the military brass.

Most of the new business deals emerged via Magic's numerous relationships. In many ways, because of his track record and brand recognition, which we'd put in the time to develop and build, it was much easier to secure these partnerships than it otherwise would've been. Magic taught me a lot about reputation, and there's a lot to be said for one's reputation preceding one into a room. For instance, when Ollie and Ben walked onto college and university campuses, the students' interest and buy-in was all but guaranteed thanks to Ollie's and Ben's notoriety in the Black community. If some off-the-shelf soldiers had walked in off the street, the students may not have acted the same and been justifiably skeptical and suspicious.

Magic also emphasized the importance of brand equity and having a solid, consistent record of overdelivering. He told me that without those two things, no one would invest in you. On a personal level, Magic taught me how to be a businessman, specifically a Black executive. His attitude and modus operandi might be considered old-school, but Magic felt the responsibility to stand up for his ancestors, to present himself in a way that would make them proud as he grew himself into a force in business. He had experienced the pain of being stereotyped based on presuppositions about his race and background. He believed in leaving nothing to chance and encouraged me to be aware of the power I had to get out in front of potential hurdles. I needed to conduct myself commensurate with the way I wanted to be treated—tomorrow and ten years from now, not just today. The way I dressed and communicated, the way I used my words in professional settings—everything mattered.

During my time working for Magic, I had my biggest aha moment that diversity—that being different—could be positioned as a serious opportunity in corporate America and eventually in the global corporate world. And he could see me excelling by operating with that mindset. Magic had an incredible operating system, where his business served as a conduit for other companies to plug in to what he was doing. He'd offer advice, communication, and inspirational speeches, as well as publicity and promotion that helped the companies he partnered with establish credibility in Black markets. Lots of today's savvy Black personalities have adopted this business model, partnering with different brands to tout everything from fashion to food and furniture. But his model was about partnerships, not sponsorships.

Magic modeled a servant-leadership style for me. I'd describe this as his philosophy that the goal of a leader is to serve the team and the organization first. This differs from "traditional leadership," where the leader's main focus is the thriving of their company or organization. I learned that it was just as important to acknowledge and thank those working the floor as it was to hobnob with executives in fancy office suites. He had a servant-leadership style by virtue of being a leader on the court and then in the boardroom, where he leaned on others and their efforts and focused on the results over everything.

In contrast, Ollie and Ben were servant leaders by virtue of being military officers. Yet their primary leadership style was performance-based. Performance was Ben's big thing, and his father had undoubtedly drilled that into his head. That was the only avenue by which Ben could fulfill his dream of becoming a four-star general. The military would be looking for any reason to discredit him, so he couldn't afford to be careless or lazy. He'd tell me the same thing he told the Black college recruits: "Don't worry about all the scholarly people making all that noise—it's about getting the job done." Ben was motivated to excel at everything, convinced that sublime execution on any task would secure more opportunities for himself and those who'd come after him. That's what inspired him, and that's what inspires me.

As soldiers, service was a way of life for Ollie and Ben. When they traveled across the country visiting Black colleges, the outcome of their investment of time and energy was a big question mark. Nevertheless, they were committed to the mission and the students they met. Ollie and Ben trusted that, regardless of the end result, being able to uplift these young Black men was valuable in and of itself. And before they knew it, that investment would pay huge dividends.

FIVE

The Pursuit of Double Victory

Our purpose begins when preparation meets opportunity.

In addition to promoting Ollie, President Franklin D. Roosevelt had solicited the general's advice on policies pertaining to Black soldiers. One of the main reasons for this was the same reason Ollie had been promoted to general—winning the Black vote in the 1940 presidential election. Yet Ollie used the opportunity of leading the Advisory Committee on Negro Troop Policies and having the president's ear to pitch the idea of pilot training for Black soldiers. Equal opportunity meant equal access to all positions in the military, including elite roles like fighter pilot. When discussing this, the president asked who would command such a unit of Black pilots. Ollie replied, "My son."

FDR wasn't sold on this plan, but it wasn't an impossible task. A year earlier, in 1939, two Black pilots, Dale White and Chauncey Spencer, flew an airplane from Chicago to Washington, DC, to draw other Blacks' attention to aviation. And future president Truman, who was a US senator at the time, helped push legislation through Congress intended to establish six initial Civilian Pilot Training

Programs (CPTPs) for Blacks, located at different colleges and universities.[1]

On paper, the Tuskegee Institute, deep in the heart of the Jim Crow South, wasn't necessarily the best location for a CPTP. The first class of twenty students, eighteen men and two women, had to start from scratch. They mowed a field, created a runway, built a hangar and a classroom.[2] Their instructor was "Chief" Charles Anderson, one of the first Blacks in the US to earn a pilot's license. During Ben's tenure at Tuskegee, he'd befriended Anderson, who lived and taught in the area. Anderson had even flown Ben as a passenger a couple of times. After spending time with Ben, Anderson felt that Ben had tremendous skills and the potential to fly planes—yet the Army was still actively prohibiting Blacks from attending pilot training.

The CPTP track that Anderson led became so successful that the War Department officially selected Tuskegee as the site for an Army pilot training program. The facility needed better infrastructure and equipment, but lacked the necessary funding for the improvements. In 1941, an unlikely advocate appeared on the scene: First Lady Eleanor Roosevelt. There's a legendary story of her being so insistent with her husband about wanting the Black pilot training program to succeed that she traveled to Tuskegee. She was quoted as saying, "I've always heard colored people can't fly, but I see them flying around here."[3] Against the objections of the Secret Service, she flew in a plane piloted by Anderson, to show people that Black pilots were safe and trustworthy.

How shocked and amazed both whites and Blacks must have been when they picked up the newspaper the next day and saw a photograph of the First Lady smiling in the back seat of a plane with a Black pilot at the controls. One thing I learned from this story and seeing the photograph for myself is the power of visual media, which can shine a light on anything and make magic happen. Thanks to this publicity and the First Lady's commitment to the administration's financial support of the program, Tuskegee could upgrade their facilities and accommodate the CPTP program for the Black pilots and other trainees.

The request for Black pilot volunteers went out in March 1941, and respondents were accepted on a first-come, first-serve basis, with priority given to those who had already completed some training with the Civilian Aeronautics Authority (CAA) or had completed at least two years of college. The call to duty that Ollie and Ben had anticipated for these young men finally went out.

While Ollie was working behind the scenes to secure the possibility for his son and other Black soldiers to become pilots, Ben and Agatha, who'd arrived at Fort Riley a few months prior, received surprising and exciting news: the chief of the Air Corps had sent a letter indicating that Ben should be released to begin pilot training, which Ollie had approved. Ben's original application, which he'd submitted in 1935 while he was at West Point, had been rejected, but six years later, the Army's plan to establish an all-Black flying unit opened the door for a second chance to become an aviator.

However, one final obstacle had to be overcome before Ben could pursue his long-held dream of becoming a pilot: a physical examination. Sadly, something that should have been routine was anything but. Upon going for his physical examination, the doctor he was assigned to lied about the results. Ben would write: "The flight surgeon . . . did not know that the Air Corps had changed its policy of not accepting black applicants and unhesitatingly failed me by reporting a falsified history of epilepsy. . . . Headquarters immediately understood why I had failed the physical and flew me to Maxwell Field in Montgomery, Alabama, for another, which I passed with no difficulty, my epilepsy having miraculously disappeared in a few days."[4]

We will never know how many other Black soldiers, including potential pilots, had such a permanent mark on their records because of the military's past discriminatory policies. Although these policies reflected the times, it's a shame to know that staff at all levels were determined to ensure Blacks were kept in their second-class status.

Along with twelve other cadets, Ben arrived in Tuskegee in 1941. Up to that point, Black soldiers hadn't been allowed to operate heavy machinery, such as tanks and aircraft, and most of their roles were service oriented. They were cooks and diggers, doing the bulk of the manual labor. Think about the training, skill, and precision required to be a pilot, particularly a fighter pilot in combat. You had to be in peak physical condition and be a navigator, an engineer, and a communicator. Almost two decades after the 1925 War College report, military leaders still refused to believe those "inferior" Black soldiers were up to the challenge.

Upon the cadets' arrival at Tuskegee, they found abysmal accommodations. While their white counterparts dined in a hall with tablecloths and Black servers, the Black cadets ate in a dining hall with dirt floors that turned to mud on rainy days.[5]

The 99th Fighter Squadron, which, along with a few other squadrons, would become the 332nd Fighter Group—later known as the Tuskegee Airmen—was officially activated on March 19, 1941. Five months later, Ben attended the inaugural ceremony for the training of Black pilots for the US Army Air Corps, where they were addressed by Major General Walter R. Weaver, the commander of the Air Corps at Maxwell Field in Montgomery, Alabama, who said:

> *The eyes of the nation will be upon you and your success will lay the foundation for others of your race to be called upon to serve in higher branches of the military service. What Booker T. Washington stood for, especially the principles of work, attention to duty, loyalty to cause—with these in front of you, you cadets cannot help but be inspired.*[6]

How immense that responsibility must have felt. In today's America, I can't think of an equivalent level of pressure on recent college graduates or people in their twenties.

It was generally understood that after Ben successfully completed pilot training, he'd command the future Black flying unit of about fifteen thousand troops, which would become the 99th Pursuit Squadron. Of this news he noted, "Naturally, I was elated, and my father immediately approved of the Air Corps' suggestion. For the first time I saw vague possibilities for a military career that could go far beyond assignments as a professor of military science and tactics at black colleges."[7]

Ben said that if it hadn't been for the timing of World War II, he probably would've aged out and never become a military pilot or commanded the squadron. When he received orders to head to Tuskegee, he was twenty-nine—only one year younger than the maximum cutoff age for new pilot trainees. Once again, luck was on his side, and he was rewarded for his patience and perseverance in believing in his goal. When the call came in, he was ready.

Over the years, about fifteen thousand soldiers trained at Tuskegee; almost one thousand were pilots.[8] The vast majority were Black college students—the very same ones Ollie and Ben had encouraged during their tours. These two officers had seen something the military establishment hadn't: the aptitude and potential of these young men, who were more than capable of defending the United States and its allies, including as fighter pilots. Though the Army was certain the experiment would fail, those who'd signed up had something to prove. They were undeterred, seeming to have internalized Ollie's and Ben's determination to use the system to defuse the system.

The Disneyfication of the Tuskegee Airmen's story, as I describe it, doesn't even come close to reflecting what they experienced. They got the oldest, most decrepit planes that were in a constant state of disrepair. They trained twice as long as other troops and had to take the flight test repeatedly. None of the generals wanted to add them to their squadron, and no one in the military leadership wanted to deploy them. Everything was a struggle.

One story that exemplifies the level of disrespect they endured took place in 1942, when the secretary of war, Henry L. Stimson, visited Tuskegee Army Air Field (TAAF) for an official inspection. After Stimson left Tuskegee, someone in Washington, DC, requested a photograph of Ben greeting him upon his arrival, likely as a public relations stunt. Since Ben had been denied a chance to meet Stimson, there was no photograph. In Ben's autobiography, he explained the problem's solution:

> *I was called in to TAAF Headquarters and given a raincoat to put on, because it had been raining on the day of the Secretary's visit. They took my picture, superimposed it on a photograph that had actually been taken during the visit, and sent it to Washington. This phony photograph was mailed to the Pentagon to prove beyond a doubt that Lieutenant Colonel Davis had met Secretary Stimson at TAAF. It appeared in many newspapers.*[9]

If you seek out this picture online or in Ben's book, you'll see that if it were taken today, it would land on a "Photoshop fail" list. It's sloppy and ridiculous-looking and demonstrates that the military was more concerned about appearances than reality. It makes me think about how strong and pervasive systemic racism was back then. Ben was an officer who was honorably serving and ready to die for the entire nation, yet was deemed unworthy of a courtesy meet and greet.

Assuredly, there were those who eagerly awaited the moment Ben crashed and burned, whether literally or figuratively. Many were waiting for their biases to be confirmed. The Army's Department of the Media, although less formal than we might imagine it today, followed him around, to ensure any errors in his judgment were documented in a video or photograph. It's like he was being hounded by paparazzi, while trying to lead a program that most didn't believe in and command troops no one wanted to get photographed with.

When I started delving into my family story, I discovered that tons of photos and hours of video had been taken of Ben both training and flying. The media wanted to capture the very moment his plane crashed, so they could splash headlines and photos across the country proclaiming that the Black pilots had failed. He was hyperaware of the scrutiny and made the Airmen aware too. Ben demanded nothing less than perfection from the soldiers he commanded. He understood how much was at stake and that failure wasn't an option. He once told me, "When you are Black, you may only get one shot. That's the reality."

When I started researching the life and times of the Invisible Generals, I discovered that in the late 1800s and early 1900s, Black newspapers thrived. Before this, I didn't even know that Black newspapers existed. Since the first one was founded prior to the Civil War, over three thousand Black newspapers in the US have chronicled the Black experience in America.[10] These outlets lauded the Black community's progress; however, Americans rarely read them, took them seriously, or elevated them to the same level as their "mainstream" papers. Day in and day out, whites consumed media focused on white accomplishments only. Additionally, whites printed negative portrayals of Blacks that reinforced stereotypes of the race as ignorant and unworthy of equality with whites. This finding made me sick to think about it and was one of the reasons I wanted to work in media.

Recently, I was contacted by the Schomburg Center for Research in Black Culture, housed within the New York Public Library. They've opened an image gallery for Ben, and I've been told that Ben and Ollie were among the top people mentioned in Black newspapers, alongside people like Jackie Robinson, Lena Horne, and Sidney Poitier. When Ollie and Ben were ascending through the military ranks, the Black newspapers tracked their every move.

At least to Black Americans, they weren't invisible at all. Though the father-son duo may have been ahead of their time for white society, the Black community waited with bated breath to witness the day when these men would finally receive the recognition they deserved from the nation as a whole.

Since Ollie flew under the general public's radar, his achievements didn't do much to alter public perceptions of Blacks in the military. Instead, the narrative began to shift when the spotlight shone on Ben and the Tuskegee Airmen. The *Pittsburgh Courier*, a prominent Black newspaper, promoted the concept of "Double Victory." The campaign "became a rallying cry for black journalists, activists and citizens to secure both victory over fascism abroad during World War II and victory over racism at home."[11]

Ben had to lobby and struggle for everything for his troops, whether livable facilities, functioning equipment and aircraft, or respect. He was the first and only Black commander in rooms where decisions were being made that could impact the entire world's fate—and his own fate and that of many Black Americans. Since he wore his race on his face, he couldn't afford to be anything less than the best, smartest, and most patient person in the room.

Having segregated units was inefficient and foolish enough, yet the absurdity went beyond the personnel. Even the maps were segregated. The base overseas where Ben's troops were stationed in World War II, which served as the only Black military base, wasn't on many of the Allied maps. This is the way my dad tells the story: Ben was concerned that if there was no clear indicator that the base was part of the Allied forces, there was a high probability they'd be attacked by friendly fire. Therefore, when they received their new P-51s, Ben ordered his troops to paint all the plane tails red, the only color they had on hand in quantity. The vibrant hue would allow their aircraft to be easily distinguished both on the ground and in the air. All these years later, you can see how the Red Tails is also a lesson in branding: simple, effective, and memorable, all while sending a message.

Although all the bomber escort units during the war had tails painted with different colors and patterns, I think it was destiny that Ben's squadron painted their tails red during the Fourth of July holiday, before their first mission in the new planes on July 6, 1944, which would be their longest-distance mission yet. Ben considered himself an American first and foremost. He felt that the opportunities he'd had throughout his life were only to be had in America. Ben believed in American exceptionalism and was highly conscious of the nation's symbols, and symbolism as well.

Along with painting the tails, pilots decorated their plane's nose, usually with a wife's name, a pinup girl, or a fierce animal like a shark or tiger. "Nose art" was a small way the pilots could express their creativity and individuality and pay homage to something they felt they were fighting for or representing. My dad *loves* telling the story of Ben's nose art—so much that he shared it in his eulogy at Ben's funeral.

In one picture of Ben, you can see the nose of his airplane behind him, with the words "By Request" painted in artistic lettering. That phrase served as a double entendre. The first meaning referred to a moment in the war when a critical mission to bomb Berlin was being planned. A list of fighter squadrons was presented, and one of the bomber commanders asked why the Red Tails were excluded. No one could answer. The commander then requested that the Red Tails escort his bombers. Ben proudly informed his men that they weren't going on that mission because they were ordered to go—they were going "by request." My dad explained it best: "By Request" was "intended to be a daily reminder to the pilots in his command that they had achieved the highest honor through their performance. Because when the chips were down, the bomber crews chose them over all the other white fighter pilots to guard them when the danger was great."[12]

The second meaning was intended for Ben's higher-ups, even if the message went over most of their heads. For many reasons, he couldn't tell them what "By Request" conveyed: they'd shown their cards by allowing their bigotry to discredit him and his pilots, instead of making decisions based on their immaculate record.

During World War II, the Tuskegee Airmen primarily were referred to as the Red Tails (along with some uncomplimentary names). Others called them the Red Tail Angels because they were responsible for escorting bombers in the air. Ben didn't tolerate any form of shenanigans, distractions, or pursuits of individual glory. He demanded his men stick to their purpose and protect the bombers at all cost, which his men did better than any other aerial squadron. Ben wrote of their deeds with pride: "Complimentary remarks from pilots, navigators, and other bomber crew members came to us by teletype or telephone. As consciousness of the job we were doing grew, crews were quick to voice their praise of the Red Tails, as we had come to be known from the painted tails of our P-51s."[13]

Ben didn't feel comfortable flying and leading every mission, although he might have preferred that. If he went down in battle, the entire program and the dreams and ambitions of so many would likely go down with him. He was the West Point graduate; he was the commander. He also needed maximum time to train and mentor the next round of Black pilots and leaders, hoping this would lead to better prospects for each of them after the war.

The mainstream media didn't cover the 99th much because the story was unequivocally positive about Blacks and their abilities. Their accomplishments ran counter to the narrative of Black inferiority that was sadly common and expected during that time in American history. However, Black newspapers eagerly covered their every move, as they always had. Throughout the war, several correspondents from publications across the country embedded with Ben and his troops on the front lines, enduring the same wretched living conditions. Ben held these journalists in high esteem and well understood their role and importance in documenting events for both the present and posterity.

An element of the Tuskegee Airmen story that's often overlooked is that the group wasn't just the pilots and ground crews. The pilots are considered the starting lineup of players, but the Airmen community encompassed all of the staff, including women and white personnel, with women typically serving as nurses and teachers. They were further yoked to the Tuskegee program when some of them married Airmen.

In 2012, the year *Red Tails* premiered, CNN published a beautifully written article about the women, detailing their involvement: "They were nurses, mechanics, supply pilots and secretaries. They nursed injured bodies and souls, packaged and repackaged parachutes, cleared land for runways and base buildings, delivered supplies and did the other work that helped keep the base running."[14] Their contributions have often gone unnoticed, so I'd like to express appreciation for and applaud their commitment to the cause. I love this quote from one of the most well-known Tuskegee women, Irma "Pete" Dryden, who married pilot Charles Dryden: "'We knew we had a purpose,' Dryden said. 'We had to make this thing work, with a passion that other units didn't have to exhibit.'"[15]

In recent years, more information has been published about the women's importance to the Airmen's success, and I know Ben would've been thrilled about that. He complimented and elevated the Tuskegee women as much as he did the men because he witnessed their effectiveness in the military, particularly during wartime, and recognized their value. He trusted and respected all women, especially Black women, and many times felt for their position in both society and the military. I have no doubt that his marriage to Agatha—his independent and intelligent companion—positively influenced his estimation of women and their abilities.

Ben would carry these forward-thinking ideas of integration into the next phase of his military career, when the United States Air Force became a separate military branch in 1947.

SIX

Founding Fathers

The world doesn't need you to live someone else's life—it needs you to be the first and only you.

In 1946, President Truman formed the President's Committee on Civil Rights, to create a plan to integrate the military. The committee provided many recommendations, yet Congress would never act on them because many Southern Democrats, known as "Dixiecrats," were opposed to integration. So Truman took matters into his own hands and signed Executive Order 9981.

When a US president signs an executive order, there's usually a lot of fanfare and photo ops with people who advised him on the directive. On July 26, 1948, when President Harry S. Truman signed Executive Order 9981—the groundbreaking directive that would ensure equal treatment and opportunity for all military service members—one crucial person was missing from the crowd surrounding him: Ollie himself, General Benjamin O. Davis Sr.

This event, a monumental occasion in American history, should have been the pinnacle of Ollie's storied fifty-year career. Instead, his voice was silenced, and his critical role in this historic moment was rendered invisible. How disappointed he must have been. Ben too.

They just couldn't seem to turn the corner on being invisible. They were in the rooms, influencing major decisions impacting the entire nation, yet they were erased at every turn. They were eminently patient and loyal, in a way I'm not sure I could ever be. As Black Americans, they didn't expect anything. They didn't want to get their hopes up because they'd been let down so many times.

Since Ollie was the only Black general in the Army—as well as one of the oldest and longest-serving military officials—his involvement in the discussion was a no-brainer. The white-majority military brass undoubtedly had their own ideas about how integration should be carried out among the ranks of the enlisted, but Ollie could speak from his lived experience of having served in a segregated, second-class military for half a century.

Regardless of Ollie's daily reality, he never lost faith in his country and its potential. Ben wrote admiringly of his father's attitude:

> *My father believed strongly in America despite all its deficiencies. A serious student of American history, he was proud of the performance of blacks in America's wars from the Revolution on. He believed that only through the full integration of blacks into society could American life be fully representative of the democratic principles so clearly enunciated by our Constitution but so poorly observed in practice.*[1]

Contemporary accounts indicate that Ollie led the commission's efforts, helping to draft the integration plan and craft the language used in the executive order. The expectation was that Truman, with Ollie at his side, would make the grand announcement of the military's leap into a fairer, race-blind future. Instead, Truman made a different announcement: Ollie would be retiring from the United

States Army. That day had been a long time coming, since the military was still resistant to allowing a Black man to ascend any higher in the ranks than Ollie already had. And the downtime between the end of World War II and the beginning of the Korean War presented an opportune time to send Ollie into retirement.

Ollie's retirement ceremony took place on July 20, 1948, with Truman himself presiding. And just six days later, on July 26, the president put pen to paper on Executive Order 9981. Ollie, having formally exited the service, received no credit for the words on the page. Nor was he mentioned in any media coverage of the event.

If Truman had wanted to honor Ollie and all of his contributions, he should have allowed the man to participate and celebrate. To give him an opportunity to stand up there and say, *This is the culmination of my career. This is why I served our nation for fifty years in a segregated military—so I could provide a better future for those who serve after me.*

My cynical side is deeply suspicious of the timing of all these events. In my research about the different presidents Ollie and Ben were involved with, I've learned that Truman was no saint when it came to race relations. According to the Truman Library website, he "was known to have the prejudices of his community when it came to views of race. He used racial slurs, told racist jokes, opposed sit-ins and intermarriage and called Dr. Martin Luther King a troublemaker."[2] The year he signed the executive order was also an election year—an election the pollsters predicted he would lose big-time. Desegregating the military would be a surefire strategy to shore up the Black vote. I tend to think the situation is very cavalier too. Truman was proud of "making Blacks equal" in the US military. But he didn't make them equal—they were already equal. I get a bit upset when people speak of "making" someone "equal."

So Ollie didn't get the credit he deserved, but his labors did greatly benefit Ben, who continued to carry the mantle of Black excellence—the ability to operate with grace in a system designed and developed to make it more challenging for Blacks to succeed—and carry on his father's legacy.

Despite the unfair world Ollie and Ben lived in, the one where the scales were forever tipped against them, they chose to make the best of it. Ben and his father both made the conscious decision to show up and be excellent, regardless of what they had to walk through, personally or professionally.

The Air Force Integrates First

As is often the case with major change in a huge organization, all of the military branches wouldn't be fully integrated for many years. The first branch to fully integrate was the Air Force, which until 1947 had been called the Army Air Forces and wasn't a separate branch. The Air Force announced they were integrating in April 1948, a few months before Truman signed Executive Order 9981 and one year after Jackie Robinson broke the Major League Baseball color barrier. And I'm proud to say that a major reason for that was Ben. His performance and that of his troops had already demonstrated that Blacks could perform just as well as, if not better than, their white counterparts. Ben himself became directly involved with integration efforts in 1948, and he felt it was one of the most important tasks he undertook in his entire military career. He provided the following insight into what he'd learned from his father and Mother Sadie:

> *My own views on integration grew out of my personal experience and the teachings I had received as a child. It had been*

indelibly impressed on me that racial discrimination was morally wrong. Furthermore, it was a cancer on the military. It continues to constitute the major problem facing our nation, and one that must be dealt with at the earliest possible time. It will not go away on its own.[3]

As Ben noted in his autobiography, both whites and Blacks opposed integration. Whites held false beliefs that Blacks were incapable of serving in certain roles. Blacks felt they'd lose out on opportunities if they were in direct competition with whites who wouldn't play fair with them. They were also concerned that integrated units would allow whites to antagonize them—something they were more shielded from in their segregated units.

The US military was the nation's largest employer. Therefore, ensuring that everyone received equal pay for equal work, and that everyone had equal access to opportunities, would set the tone and expectations for the rest of society. The military was intended to represent America in all its diversity and potential, and if an organization as vast and complex as the military could successfully integrate, there was zero reason every other industry couldn't do the same. Ben noted, "These policy changes in the Air Force placed it far out in front of existing social relationships in the United States. They were implemented in spite of the strongly voiced objections and misgivings of senior generals. . . . There was the deep fear that moving ahead of the nation's practices in racial relationships would almost certainly adversely affect morale and efficiency."[4] Thankfully, all those concerns were unfounded: "Most people now agree that the Air Force became a far more effective organization as a result of integration. For the first time, it was able to train, assign, and deploy its black manpower with complete flexibility. . . . The Air Force's new policy was a first and important step, but it would take many years before it was fully implemented throughout the armed services."[5]

As John L. Frisbee noted in his book *Makers of the United States Air Force*, which profiles twelve men who are considered founding fathers of the military branch: "Segregation eventually would have been banned in the armed forces had there not been a Ben Davis. It is a reasonable assumption, however, that racial integration in the military would have followed a different and probably much slower course toward equal opportunity had it not been for him."[6]

A Father Figure Inspires His Men to Achieve

I love learning new things about Ben—especially hearing firsthand accounts from people who knew him personally. It makes me feel closer to him, even over two decades after his death.

Recently, I read about an example of buried Black achievement, which the media refers to as a "Top Gun" competition (not to be confused with the Navy's TOPGUN program); its formal name was the United States Air Force First Aerial Gunnery Competition.[7] When the meet was announced in 1949, four Tuskegee Airmen were chosen to participate in Las Vegas. Their commander, none other than General Benjamin O. Davis Jr., met with them prior to their departure and said, "If you don't win, don't come back."[8] Ben's men greatly respected him, and some of them considered him not only a leader but also a father figure.

Even after the Tuskegee Airmen's successful combat experience in World War II, people remained dubious about Black pilots' skills. Thus, no one expected them to come close to winning this competition. That they were flying older, less advanced aircraft made that even more improbable. Yet as Lieutenant Colonel James Harvey III, one of the participants, asserted, "It's the skill of the pilot that determines what's gonna happen. They were there to compete, and we were there to win."[9] Against all odds, they won. A February 2022 news article described the scene:

> *But when their names were announced, the room remained quiet. There was no applause. A photographer snapped a single photo of the team with its trophy, which was left in storage for 55 years afterward.*
>
> *"Our victory was swept under the rug," retired Lt. Col. James Harvey III, former fighter pilot with the 332nd Fighter Group, told* USA TODAY.[10]

In 2004, a historian discovered the trophy and fought to have it displayed in the National Museum of the US Air Force, and in 2022, a plaque was placed at Nellis Air Force Base commemorating the pilots' victory at the site, which is right outside Las Vegas. It took almost three-quarters of a century for these men to receive the recognition they'd earned.

Regarding the import of the moment, Harvey said, "If you believe in something and set off on a mission to correct something that was wrong, don't give up. . . . Stay with it. We waited for this 73 years, and it finally paid off. Set your sights high, always."[11]

How many stories like that of the victorious "Top Gun" pilots remain lost to the past? How many heroes are we lacking, whose example could inspire and improve us as individuals, and as a nation?

Ben fully understood the need for youth, particularly young Black men, to have people to look up to. In 1953, he created an Air Force recruitment tool that I consider yet another of his overlooked contributions: the Flying Thunderbirds, an air demonstration squadron. Interestingly, in 1956, the Thunderbirds were relocated to Nellis Air Force Base—the same place Ben's men had won the "Top Gun" competition several years prior. Today, these pilots still perform at air shows across the country, allowing their amazing aerial talents to wow audiences and inspire the next generation of military pilots.

Forging a Family

Amid all of Ben's professional achievements, his personal life lacked something that he and Agatha both longed for: a child. But he and Agatha never had a child of their own.

While Ben and Agatha agonized over their inability to have a child, her sister Vivienne and her husband, Lawrence, didn't want any children. She'd thus terminated several pregnancies at different points in her life, and when she became pregnant an eighth time, she had a long discussion with Lawrence about not carrying the baby to term. Her doctor, who was a family friend, advised her that due to the number of previous procedures she'd had, it would be in her best interests to give birth to the baby; another terminated pregnancy could cause serious medical complications. At that time in American history, abortion was illegal, so the procedure varied in regard to safety. With the physical reality and her doctor's advice in mind, Vivienne decided to continue her pregnancy, and Larry Scott Melville—my dad—was born on October 25, 1933.

Not long after Larry was born, his parents drove him from the Bronx to New Haven, Connecticut, and dropped him off at his grandparents' house. His grandparents would raise him for the first six years of his life. He had minimal contact with his parents, who only visited him occasionally on the weekend. Agatha often stayed with her parents during these years, while Ben was deployed. She took over most of the childcare duties, to relieve her aging parents of the responsibility.

She'd go for walks with Larry down the street and in town, and whenever anyone asked if he was her baby, she'd say, "Yes—he's mine." She and Vivienne were very close and had agreed to share the child. As far as Agatha was concerned, since she was there and Larry's mother wasn't, he was *her* baby. She was filling the void her and Ben's childlessness created in their otherwise fulfilling lives.

When he was six years old, Vivienne showed up in New Haven one weekend and said she was taking him to live in New York City, because she wanted him to attend school in New York, even though he was already enrolled in school in New Haven. She snatched him away from the people he loved and the only place he knew as home.

Not surprisingly, after the move to NYC, Larry started acting out. He refused to cooperate with anything his parents asked him to do and was quite belligerent. Perhaps in his child mind, he thought that if he made his parents' lives miserable enough, they'd give up and send him back to New Haven, which is what he wanted.

Vivienne and Lawrence traveled a lot, and as Larry got older, he desired to spend a whole year somewhere. The most logical place was Tuskegee with Ben and Agatha. Vivienne was adamant that he remain in school in NYC, but Agatha was ready to fight for her and Ben to take Larry.

The details of the actual day and way Larry moved to Tuskegee are unclear, but somehow, Agatha maneuvered him away. And the result was that he ended up staying with them much longer than a year.

The Black family dynamic during this period, similar to today, could be complex. Particularly among family members, legal adoption wasn't the way. Children were raised in more of a "village environment"; whomever had the means and desire to raise a child would do so, whether it was grandparents, aunts and uncles, or a close family friend. No formal paperwork links Larry to Ben and Agatha—he was always their nephew. Even so, they raised Larry as their son, as someone who could continue the Davis family line and their traditions.

Not surprisingly, with his military background, Ben was a strict disciplinarian—much as his father, Ollie, had been. The importance of education was instilled in Larry. He was told that without an education and a clean record, there were neither hope nor options. Education was the key to opportunity. Being surrounded by Black excellence at the Tuskegee Institute inspired young Larry. He'd hang out with the likes of George Washington Carver, the famous botanist who worked at Tuskegee for over forty years, as well as some of the best and brightest intellects in the Black community.

Maybe thanks to his immersion in this academic environment, Larry was a scholarly young man. As he progressed in his education, he didn't know what he wanted to do with his life. Ben encouraged him to carry on the family mission of changing and improving America through service—to look to the future and see where he could contribute and have the greatest impact.

Thanks to Ben's influence, Larry developed a passion for aviation and worked every day to improve his skills and knowledge so he might become a military pilot, following in Ben's footsteps. Unfortunately, all the training in the world couldn't compensate for the poor eyesight that put Larry's dream off forever. Back then, you had to have 20/20 vision to be an aviator, or else the military considered you a liability. Since my dad needed glasses to see clearly, he was automatically disqualified from becoming a military pilot.

With Larry's dream of becoming a military pilot off the table, he began building and flying model airplanes. Ben encouraged this hobby, while seeking options for a sustainable career for his nephew. One day he made the fateful decision to put Larry on a plane to Santa Monica, California. He'd meet and spend time visiting with Marcus O. Tucker Jr., a family friend who was attending law school at Howard University. Marcus said that Larry should consider the legal field, where so much was happening around civil rights in the late 1950s and '60s. Though Larry wasn't yet ready to make that commitment, the seed had been planted.

In the interim, Larry was drafted to serve in the Korean War. Ben was concerned that Larry wouldn't survive if he ended up in combat. And it would break his and Agatha's hearts to lose him. Ben advised Larry to request a clerical job, where he could develop skills transferable to the workforce after his deployment—typing; working with people from various walks of life; being part of a team; dealing with bureaucracy, policies, and procedures; and understanding discrimination.

Ben wanted Larry to be educated about the realities Blacks faced in different parts of the country, particularly the way Jim Crow laws in the South impacted Blacks' daily lives. Since Larry had spent his early years in the North, he'd never experienced segregation up close and personal until he went to Tuskegee. It was jarring. To this day he still has some of the photographs he took when he first saw signs that said "Whites Only," "Colored Only," and "No Colored Allowed." He clearly remembers Ben getting upset with him for going off the base and interacting with locals, in what Ben referred to as "a sea of angry whites." Ben didn't drive at night, which Larry found odd, but he refrained from doing so because he was warned that the local authorities told Black people that if they left the base "they'd get pulled over," which in those days was code for "you might not return home." Larry quickly fell in line and only left the base to explore the airfield and the Tuskegee Institute.

Later, I'd find out that these experiences drove my dad to work in the legal field, to help abolish these discriminatory laws. Similar to Ben's moment of enlightenment when he took his first flight with the barnstormer, and Ollie's inspiration to become a military officer after being a cadet in high school and seeing the Buffalo Soldiers, Larry saw segregation as an evil sickness that affected everyone—Blacks and whites alike. His goal wasn't to march or fight the power, but to use the system to defuse the system.

Regarding his deployment, Larry wanted to reach out to Ollie to help ensure that he'd get a noncombat position, but Agatha told him not to do that. I can't help but wonder if Ben pulled any strings to keep my dad off the front lines. If he did, that's just one more reason I can be grateful to him for looking out for my dad and preserving his life.

After being discharged from the Army, Larry applied to the law school at Howard University and began his studies in 1959. There, he was pleased to continue his friendship with Marcus Tucker, who'd been instrumental in his decision to attend Howard.

It was a thrilling time to be a law student at Howard, a tight-knit community with fewer than two hundred students in the entire law school. The program was home to future heroes of the civil rights movement like Vernon Jordan Jr., who graduated a couple of years ahead of Larry.

Future Supreme Court justice Thurgood Marshall often utilized the students' presence and minds to help him practice for cases he'd be presenting, placing Larry in the epicenter of civil rights law. Marshall operated similarly to Ollie and Ben, working within the system to evolve the system and with faith that America would one day fulfill its promise of equality for all, not just on paper but also in practice. Larry saw that the legal field could be an effective avenue to enact change.

Upon graduation in 1962, Larry moved back to Connecticut with the intent to start his own legal firm, and he immediately encountered obstacles. Due to his race, no one would lease him a space to use as his office. At that point, he had a much deeper knowledge of segregation after having lived in the South. When he dealt with it after returning to the North, his attitude was more one of acceptance and strategizing how to work around it versus getting caught up in it. It's the same philosophy I have today as a chief diversity officer, and a way of thinking that has been passed down in my family. A mindset that achieving any goal may not be easy, but it's important to lock in on a goal nonetheless, to help you stay focused and better enable you to not get caught up in the noise. While sometimes the noise follows you and is impossible to avoid, this mindset is about the importance of having clarity, conviction, and confidence as you chart a path forward for your life. This mentality is probably why my dad didn't want me to pursue a career in diversity. It's acceptable and more widespread now, but back then he was upset and concerned that "getting caught up in race matters" at work was the fastest way to kill your career and future prospects.

Ben helped Larry navigate that situation by suggesting that they put the paperwork for the lease in his wife Sonja's name. Although my mom is biracial, she can pass as white, and no one questioned her about her race. Once Larry was settled, he began his law career in earnest, frequently doing pro bono work.

My dad never told me much about his work, including specific cases. In 2018, I gained a bit of insight when someone shared with me a copy of a book titled *Murder in the Model City*, which was published in 2006. My dad was a special public defender for a member of the Black Panther Party, whose freedom he was able to secure. The book even includes a picture of my dad with his client, and the trial received national coverage.

I gave my dad a copy of the book, and his reaction was a mixture of disbelief and confusion. He asked, "Why would they make a book about this?" Like his forebears, he moved through the world with humility and a servant-leader mindset.

Another example of my dad's humility relates to a mysterious photo he had tacked to the corkboard in his chambers, which had been there as long as I could remember. In the picture, he's standing with an older gentleman. For most of my life, I didn't give the picture much thought. But one day, it occurred to me to ask him about it.

"Oh, that was me and George Washington Carver," he said with a shrug.

"George Washington Carver? Like *the* George Washington Carver?" I said, incredulous.

"Yeah. When we lived in Tuskegee, Ben and Agatha used to drop me off at his house sometimes." His nonchalance let me know he didn't plan to tell me anymore about it.

Similar to his appearance in *Murder in the Model City*, my dad downplayed the unique and extraordinary achievements and experiences he had while living in Tuskegee. There's little doubt in my mind that Ben had modeled this approach for my dad: don't draw attention to yourself—working quietly behind the scenes will keep you in the game longer and allow you to accomplish more. This understanding of the power and perception of invisibility, and the ability to leverage it, proved to be a double-edged sword and a complex reality for Ollie, Ben, and my dad during their professional and personal endeavors. Invisibility was deployed as a superpower or secret weapon when they wanted to deflect attention. But as their careers continued, an unintended consequence appeared: the system they chose to move in was unaware of these achievements, and therefore, many people didn't know about these accomplishments either. The invisibility that ensured their survival in the short term set them on a path toward their achievements being erased in the long term. Invisibility was both their oxygen and their pain.

Over the years, Larry got involved in politics, and he'd leave his mark on the Connecticut legal system, when he became one of the state's first Black superior court judges in 1978. Making the move to judge meant he'd earn less than half his income as a lawyer, but he felt he couldn't place a dollar value on the influence his new role granted.

I wouldn't learn many of these details about my father's life until I started on my journey of discovering the Invisible Generals.

Throughout my life, I'd wondered why and how my dad had ended up living with Ben and Agatha. It was an unspoken reality in my family. Honestly, it never occurred to me to just ask—and until recently, I'm not sure how forthcoming my parents would've been with this information. However, as they've gotten older and their time on earth is getting shorter, I think they realize that if they don't share these stories now, they'll be gone forever. I appreciate that they've decided to open up more, even though I know sharing some of these intimate details has been hard and has brought up painful memories.

Over the years, I've come to appreciate the many ways Ben and my dad leaned on each other throughout their lives. They fostered an unbreakable father-son bond that lasted almost seventy years, until Ben's death. In the nurturing, stable home that Ben and Agatha raised Larry, he had access to the generational collateral that would provide a foundation for his future extraordinary achievements. My dad learned from the best what it means to help shape a more equitable world that values everyone's contributions. And I'm still learning from both of them.

SEVEN

Look Forward, Not Down

Even when the path forward is unclear, we can gain clarity along the way.

By the time Ben retired from the military in 1970, it had been almost thirty-eight years since he'd reported for duty at West Point. During his career, he had worked under six US presidents and commanded troops in three major wars: World War II, the Korean War, and the Vietnam War. He'd been a model soldier, done everything that was asked of him without complaint, and accomplished much. In spite of all he achieved, he was denied receiving his fourth star, an honor no Black had ever earned. This level of achievement threatened not only the powers in charge but also the system that had built its strength on the narratives of Black inferiority. Ben was indeed too good, and being passed up for this recognition reinforced the unspoken reality that there were those who felt that Blacks needed to be "kept in their place."

After World War II, especially with the success of the Tuskegee Airmen, Ben was becoming a household name and ascending to a position of power and influence, in both military and civilian realms. I even have a newspaper clipping with an article titled "America's First Negro Four-Star General?," which indicates that he was on people's radars. He was supposed to receive his fourth star in 1967 and was more than qualified for that accolade. The paperwork was on President Lyndon B. Johnson's desk, but Johnson believed he'd done enough for civil rights, having taken up the cause after President Kennedy's assassination in 1963. Johnson had signed the Civil Rights Act in 1964 and nominated Thurgood Marshall to the Supreme Court. Johnson may have felt he couldn't garner any additional political gains by promoting Ben.

When it became evident that Ben wasn't going to get the fourth star, he recognized that, like his father before him, he'd be pushed out of the military without having reached one of his ultimate goals. For Ollie, the goal had been desegregating the military, which Truman did after forcing the elder General Davis to retire. And here was history repeating itself, with Ben denied becoming the first Black four-star general in the US military.

The disappointment of your best not being good enough can be devastating, yet Ben Jr. used this experience to propel his postretirement life into a dynamic public sector career. His preference was to continue working in aviation—even if that meant flying a commercial airliner or for some kind of delivery firm, like the US Postal Service or the newly founded Federal Express (FedEx). Though the Civil Rights Act was passed several years before Ben's retirement, many companies violated the law by still refusing to hire Blacks. The airlines were no exception, though American Airlines led the way when they hired the first Black commercial pilot in 1964.[1] Not much has changed since then—even today, 93 percent of US pilots are white men.[2]

Prior to his retirement, Ben received a firm job offer from the assistant secretary of defense, "for a job dealing with racial matters in the Department of Defense," which he declined: "I turned it down because it did not carry enough authority for me to be effective, and I did not like the idea of being chosen for a job on the basis of my race."[3] He also turned down several other offers, such as a professorship at the University of South Florida and an unspecified job at Pepperdine University. The Pepperdine job had an annual salary of $15,000, and Ben once told me flat out, "It was an insult."

Ben's first job after retirement from the US military was as a director of public safety in Cleveland, Ohio—a city where he'd attended high school and a place he knew and loved. Although the prospect of partnering with Carl Stokes, the first Black mayor of a major US city, appealed to him, Ben only lasted several months in that role. He succeeded in helping reduce the city's crime rate, but in the end, he felt this particular city and state government was a political minefield and didn't want to contend with all of the drama.

Ben and Agatha, who'd been married over thirty years and traveled the world together, relocated to the Washington, DC, area, where they felt most comfortable. But DC was missing one thing Ben loved: the sight and sound of airplanes, so they settled across the Potomac River in Arlington, Virginia, in an apartment with a balcony that not only overlooked the Washington Monument but also allowed Ben to see and hear the air traffic of Washington National Airport (now named Ronald Reagan Washington National Airport).

Beyond his happy place in Arlington, he enjoyed his freedom and exploring opportunities to continue serving the United States. And 1970, his first year of retirement, would prove eventful for both personal and professional reasons.

Prior to resigning from his director job in Cleveland, he joined President Nixon's Campus Unrest Commission, which was established in response to campus unrest fomented by Vietnam War protests, racial strife, and other issues afflicting America's colleges and universities. At both Kent State in Ohio and Jackson State in Mississippi, students had been shot and killed by National Guardsmen. Ever striving to keep his fellow citizens safe, Ben wrote, "I was happy with the assignment because of the nature of the problem and the people I would be working with. The tasks outlined by the President for the commission were challenging and vitally important for the nation."[4]

After entertaining a few new employment options, Ben returned to DC area and entered the sector his father had long ago shunned, yet was the place where his grandfather Louis had built the Davis name: the federal government. He held a few different positions over the course of the next eight years, and in those roles his influence and importance never diminished. He'd flourish and continually demonstrate his innovative and entrepreneurial spirit, as well as his patience and perseverance in an environment that might accept him for who he was—a pioneer.

When Ben returned to the Washington, DC, area, he met with President Nixon, Attorney General John Mitchell, and FBI director J. Edgar Hoover. He was tapped to consult on aviation security, and within a few months' time this evolved into the role of director of Civil Aviation Security. Ben would report directly to the secretary of transportation and work in concert with the Federal Aviation Administration (FAA). A year after Ben began working in DC, Nixon nominated him for assistant secretary of transportation for safety and consumer affairs, and he was sworn in to the position he'd hold for the next five years.

At the end of the year, Ben suffered a tremendous personal loss when his father died on November 26, Thanksgiving Day, at the age of ninety. Ollie had been living in Chicago, with Elnora and her husband. His death was reported on the front page of the *New York Times* on November 27. On November 30, Ollie was buried in Arlington National Cemetery with full military honors. Ben wrote, "During the funeral I thought about my long and happy relationship with this remarkable man. I had always given him my full respect, and I do not recall any serious disagreements or harsh words of any kind between us.... To my surprise, I found myself overcome with grief, and I am experiencing the same feelings as I write these words two decades later."[5] The power of the father-son bond between Ollie and Ben remained strong until Ben's death, and I never heard him say anything negative about his father, even though my dad had told me that Ollie could be a hard man—a strict disciplinarian, micromanager, and someone who always colored within the lines. As Ollie worked his way up through the ranks, he recognized his privileged status, while also acknowledging that, as a soldier, nothing was guaranteed—even life itself. He also knew his presence in the military was a problem for some, so he operated by the book 100 percent, with every move deliberate, focused, and goal-oriented.

Other than Ben's desire to work in aviation, he was invested in keeping America and its citizens safe, which was one of the reasons he'd taken the Cleveland job. Though his efforts there proved futile because of the political climate and local government maneuvering to involve him with issues he wasn't in a position to address, he met with much greater success at the Department of Transportation.

His position was created in response to an uptick in skyjackings. When most contemporary Americans think of skyjackings, the tragedy on September 11, 2001, is probably the first thing that comes to mind. However, skyjackings were once more commonplace, and the US experienced about two hundred skyjackings or attempted skyjackings in the 1960s, on flights between the United States and Cuba.[6] The frequency had started to rise in 1970, so they recruited Ben to tackle the issue.

Ben started researching solutions and was shocked to discover that only two airports in the entire country had security—National and Dulles, both federally owned and operated.[7] Addressing the problem would require an immense amount of communication and cooperation with all domestic airports. To combat the skyjackings, one of his first strategies was to implement a commercial airport security screening process that used both low-dose X-ray scanners and metal detectors, to identify foreign and potentially harmful objects that passengers were attempting to bring on a plane, whether in their luggage or on their person. This whole system that Ben created evolved into what we know today as the Transportation Security Administration (TSA), the organization formed in response to September 11.

I was able to locate an incredible photo of Ben walking through the first metal detector they tested, slightly hunching his six-foot-two frame to walk through the machine without hitting his head on the top. The side of the photo is dated January 11, 1971, and a note indicates the photo was taken at a commercial airport security conference. After the system proved effective, it was implemented both in the US and abroad. The results speak for themselves, as Ben wrote: "Worldwide, 22 hijackings occurred in 1973 and 19 in 1974, but the 50 states were free of either a hijacking or an aircraft bombing. Skyjacking in the United States had become a thing of the past."[8]

A second strategy Ben identified to increase airline security is what became the Federal Air Marshal Service. At that point, onboard security or law enforcement was handled on a state-by-state basis, and he converted these endeavors into a federal program, to ensure greater consistency and efficiency. Working with FAA chief John Shaffer, Ben oversaw the training and deployment of four thousand undercover agents. Training and expectations were rigorous: "The marshals, armed with .38-caliber pistols, would be instructed to shoot to kill; no man was supposed to qualify for the job unless he could fire twelve bullets in twenty-five seconds with enough accuracy to kill a hijacker from forty-five feet away."[9]

One unexpected event that surprised Ben in retirement came courtesy of an institution he had little to no contact with after graduation: West Point. He received a letter saying that the Washington, DC, chapter of the class organization had unanimously voted for him to receive an alumni award "For Outstanding Service and Dedication to His Country and Alma Mater." He wrote, "I had no mixed feelings about accepting the award—to refuse it would have been ungracious. Through the years several of my classmates had expressed regret about the way I had been treated at West Point, and I held no grudges."[10] It was a small bit of recognition and redemption, which Ben gratefully accepted—as he always did.

Ben retired again in 1975, but that was short-lived. The next year, he accepted a new position: special assistant to the secretary of transportation for the national maximum speed limit. His primary objective was to provide recommendations pertaining to a national speed limit, to keep drivers safe by minimizing speeding and accidents and to save gasoline by ensuring that cars drove at their peak rpm (revolutions per minute). Ben, the face and voice of the program, was nicknamed "Mr. 55." Today, even though states can set their own speed limits, almost every American can remember seeing the iconic black-and-white 55-mile-per-hour speed limit sign. Yet as commonplace as the speed limit is, most people don't know it was created by Ben.

I've learned many fascinating things about Ben from people on the internet, who find me and reach out to me with a link, a quote, or a comment about him. Someone online showed me the photograph of Ben and President Jimmy Carter unveiling the 55 mph sign in the Oval Office. It's nice to hear and see how Ben affected families and to hear from the grandchildren of those who knew him. Sometimes it feels like I'm learning something new every day. Not only because I'm more knowledgeable about where to look for information and the increased number of Google Alerts I get but also because of the many notes and emails I receive from people who want to share a Ben story with me and my family.

A hallmark of Ben's retirement was his influence on the United States Air Force Academy (USAFA). He was never one to just talk the talk, and he advocated for women in the military when that wasn't a cause du jour. I think he had immense empathy for women because of his own times of being ostracized and underestimated. So when he served as chairman of the USAFA Board of Visitors, he prioritized women and helped build out the academy's gender policies. Of this experience, he wrote:

> *At first, the Superintendent and all the board members except me opposed the entry of women, as did many other prominent military people. . . . Our lively discussion ended with the recognition that the participation of women was the wave of the future, and that they should be appointed to the Air Force Academy. The board took this position in its report.*[11]

In addition to ensuring that women could apply to the academy and become cadets, he included practical accommodations for them, with measures such as modifying the curriculum to be more inclusive and providing adequate housing and facilities on campus. Ben demonstrated that it's one thing to say you're committed to integration (perhaps today we'd use the phrase "diversity and inclusion") and another thing altogether to create an environment to support that commitment.

The first class of 157 women entered the USAFA in 1976. One of those graduates, Janet C. Wolfenbarger became the Air Force's first female general in 2012.[12] In 1993, Jeannie M. Leavitt would become the Air Force's first female fighter pilot and was later promoted to brigadier general. Then in 2000, Lieutenant Colonel Shawna Kimbrell became the Air Force's first Black female fighter pilot.[13] I had the honor of meeting Lieutenant Colonel Kimbrell in 2019. Although it's taken decades for these strides to be made, one of the reasons they're being made at all is because of Ben.

In the end, Ben's inability to work as a commercial pilot may have been a blessing in disguise, both for American citizens and Ben's sense of accomplishment. His knowledge of the military, government, and policy—along with his track record and professional demeanor—led to a second act that was as unlikely as his first. Think of how different both road and air travel would be without the ubiquitous speed limits and airline security.

Above all things, Ben loved his country and his fellow Americans and wanted to do everything he could to keep them safe. Those motivators had sustained him through four years of being silenced at West Point, seen him through combat and commanding troops in three wars, and compelled him to continue finding ways to serve even after he retired from the military.

When I'm facing adversity, especially in my career, I bring to mind the previous generations in my family. For instance, a coworker once told me, "If you were white, you'd be so much further ahead." That statement could have made me throw up my hands and say, "Why the hell should I even bother? It's not worth it!" I could have allowed despair and disillusionment to deter me. But if I were standing in a room with my tenacious forebears, what excuse could I offer them if I'd given up?

I adopted the modus operandi of those men: use the system to defuse the system. I felt confident that in the corporate world, I could scale my efforts and build critical mass in a way that would profit both the Black community and America as a whole.

In today's corporate world, "diversity" tends to focus on hiring employees from groups that were previously excluded from the candidate pool. While that's a critical part of the evolution, two other focal points are equally crucial. The first focus is *representation*, making sure the people in the commercials, advertisements, and boardrooms reflect the customer base. The second focus is ensuring variety across the supply chain and vendors—the producers, directors, videographers, photographers, and so forth.

From my seat at the table, I've been able to speak into the marketing and advertising of some of the biggest and most recognized companies in the world, like Apple and Amazon, collaborating with them and providing the advice, counsel, and story to provide the fact-based support they need to build more inclusive and diverse programs. Throughout the creative process, the goal is to invite more people into the conversation, instead of having a small, exclusive group of decision-makers walled off from the firm's more diverse staff. That strategy both improves the product or service and fosters a stronger connection to consumers.

Ben always told me, "Look forward, not down." His life epitomized this thinking. He could have allowed President Johnson's denial of his fourth star to make him bitter and resentful, but instead he seized the many opportunities presented to him, so he could continue to pave the way for those who came after him. And I hope to do the same thing along the winding road of my own career, wherever it takes me.

EIGHT

The Million-Dollar Star

Sometimes you're tolerated but not celebrated; other times, you're celebrated but not compensated.

Every time we visited Ben and Agatha, they'd discuss politics with my mom and dad ad nauseam. The same thing on their Sunday phone call every week at the exact same time, without exception. Agatha in particular enjoyed debating my dad. My mom said that time was reserved for Agatha, and it was like holding court with the Queen.

I never paid attention to or participated in these conversations because politics didn't interest me at that age. As a kid and young adult, talking politics was the quickest way to lose my interest. Sports, music, fashion, entertainment? Sure. Politics? Nope. What went on in Washington, DC, was far removed from my daily life.

When Bill Clinton ran for office in the early 1990s, the media couldn't get enough of him. It seemed as if no opponent stood a chance against this charming, funny, charismatic governor of Arkansas. And you'd be hard-pressed to find anyone my age who doesn't remember when Clinton appeared on *The Arsenio Hall Show*, wearing shades and playing the saxophone. People ate it up. For the first time in my lifetime, a presidential candidate seemed cool and in touch. After he won the 1992 election, some even referred to him as the "first Black president." Americans of every age and walk of life appeared to rally behind him.

Bill Clinton was the first president I voted for, when he ran for reelection in 1996. Thus, I had a bit of a personal attachment to him. I took pride in having done my civic duty by voting, which hadn't been guaranteed for Blacks until the Voting Rights Act of 1965 was passed under President Lyndon B. Johnson.

I was somewhat surprised to learn that Ben never voted. He loved to talk about politics with my parents, yet he felt his contribution to America would be through his performance and leadership in defending the country and its citizens. He felt that voting as a military man, and specifically as a general, could be used against him depending on the political party that was in power. He found that being nonpartisan was critical to his success because he assessed his commander in chief based on the president's merit and behavior, not political party. Not only this, but he said that behind closed doors the two major political parties had a similar agenda, so no matter who you voted for, you'd essentially end up with the same result.

One year into Clinton's second term, he became embroiled in a major scandal. In spite of his denials, the truth surfaced that he'd been having an affair with a White House intern, Monica Lewinsky.

The story exploded, and news channels breathlessly covered every minute detail of the ongoing drama. A whole cast of characters emerged—Linda Tripp, who'd recorded private phone conversations with Lewinsky, during which the intern chatted about the affair. Kenneth Starr, the independent counsel assigned to the case. The situation was rich fodder for comedians. Jay Leno joked about it daily on *The Tonight Show*, and *Saturday Night Live* featured regular skits about it. Though it was easy to laugh at the shows' humor, the gravity of the situation became evident once the impeachment trial kicked off. A US president hadn't been impeached since Andrew Johnson in 1868. Over a century later, would Clinton be next?

Amid this chaotic circus, Benjamin O. Davis Jr. would become a four-star general—sixty-two years after he graduated from West Point and twenty-eight years after he'd retired from the military.

Senator John McCain, a retired Navy vet who'd been a prisoner of war (POW) during the Vietnam War, was a passionate, tireless champion for the military and veterans. Under both Republican and Democrat presidents, McCain had spent years advocating for Ben to receive his fourth star. Ben's résumé and command over thousands of troops in multiple conflicts usually would've resulted in achieving that four-star status. McCain felt this injustice needed to be rectified.

All I knew was that my family had been invited to the White House—on the day the Christmas tree would be lit, with lots of activity and energy swirling—and that Ben would receive his fourth star. I knew next to nothing about the military then, or the historical significance of what was happening. I was just excited that my family was meeting the president.

That Ben wouldn't be the first Black four-star general did nothing to diminish my enthusiasm. One of Ben's fellow Tuskegee Airmen, Daniel "Chappy" James Jr., had achieved that status in the Air Force back in 1975. Since the Air Force was founded in 1947, only 230 out of the millions of active-duty personnel have become four-star generals. Truly rarified air. For a Black American, even more rare. Ben's promotion was actually written into the National Defense Authorization Act in June 1998, as Amendment 2993:

SEC. 531. ADVANCEMENT OF BENJAMIN O. DAVIS, JUNIOR, TO GRADE OF GENERAL.

(a) Authority.—The President is authorized to advance Benjamin O. Davis, Junior, to the grade of general on the retired list of the Air Force.

(b) Additional Benefits Not To Accrue.—An advancement of Benjamin O. Davis, Junior, to the grade of general on the retired list of the Air Force under subsection (a) shall not increase or change the compensation or benefits from the United States to which any person is now or may in the future be entitled based upon the military service of the said Benjamin O. Davis, Junior.[1]

The day before the event, when the immediate family converged on Ben and Agatha's apartment in Arlington, we learned of the two conditions for Ben's fourth star. One, he was receiving the star in retirement, not while on active duty. Which wasn't a surprise because of his age. Although other Air Force generals have gotten a fourth star in retirement, they're typically not recalled into active duty. But number two hit differently: he was relinquishing any claim to financial compensation or benefits from the promotion, from the time he should've received his fourth star to that year (1967–1998) or from that year until he died (1998–2002). The latter condition, stipulated in the amendment, caused some serious family drama and heartbreak. Under normal circumstances, the promotion to four stars would've come with increased retirement income and various perks.

Agatha was furious. She couldn't fathom that with everything our family had done for the US military and government—all the sacrifices, indignities, and adversity they'd lived through—they weren't going to pay Ben the money he was due for the honor he was receiving. She told Ben not to accept the fourth star because the whole thing was disrespectful. "It's an *insult*!" she'd shout. "*An insult!*" My dad shifted into full-blown mediating judge mode, trying to calm Agatha.

Amid her stormy rage, Ben remained as calm as ever. Although he'd been that way as long as I could remember, this tendency had become more pronounced for a specific reason: he was in the early stages of Alzheimer's disease, something few people outside the family knew. One thing I didn't know then was that Agatha, too, was struggling with cognitive decline. Their age and advancing mental disease seemed to heighten their respective personalities—Ben became more gentle, Agatha more combative.

While Agatha ranted and raved, Ben looked at his wife of sixty-two years and said, "Agatha, what difference does it make? I promised my father I'd be the greatest general in American history. Today I'm going to agree to this honor."

"Well, I'm not going. I'm not gonna go. You shouldn't agree to this," she told him.

Usually he would've heeded her complaints and said, "I understand. You're right, Agatha." But not that day. He agreed, and the paper was signed without hesitation.

Late into the evening, my mom helped Ben write his speech. They went over it at least five times, until she believed he'd committed most of it to memory. When she asked him, "Do you have it?" he replied, "Yep, I've got it."

Agatha couldn't believe that we were going along with this. All these years later, I still wonder about the atmosphere in their apartment after we left that night to go to our hotel. Did Agatha continue to argue with him and antagonize him? Or did she ice him out the way she had the rest of us?

The next morning, we again met at the apartment. Ben donned his full military uniform, looking thoroughly dignified. It was my first time seeing him in his Air Force dress blues, sharp from head to toe. Agatha was in her pajamas, and obviously still in her anger and resolute in her refusal to attend the ceremony.

"Do you need me to stay with you?" my mom asked her.

"I don't need anybody to stay with me," Agatha replied. "You, all of you, just go." Her dour demeanor couldn't dampen our spirits. We couldn't wait to celebrate this moment with Ben.

While my family waited at the White House for the ceremony to begin, we heard murmurings around the room about the timing of the event. People wondered aloud if this was an attempt to distract people and the media from the impeachment proceedings. The House of Representatives inquiry had been launched in October, and it was probable that Clinton would be impeached before the end of the year, with a high probability it would happen that week. Regardless of the timing, my family and I proudly took our seats in the first and second rows of the crowded room. I sat next to my parents in the front row. On my other side was the empty seat reserved for Agatha, her absence palpable. My brother flanked the vacant chair, which became occupied by a longtime family friend, Angelique Tucker—the daughter of the Honorable Marcus O. Tucker, who'd encouraged my dad to attend law school with him at Howard University all those years ago.

If you visit the YouTube channel for the Clinton Library, you can watch the scene unfold. A few opening remarks were delivered, then the Tuskegee Airmen presented Clinton with a red blazer and declared him an honorary Airman. Clinton took the stage and delivered a powerful speech about the character of General Davis. To this day, one of my favorite lines from Clinton's speech inspires me: "You are the very embodiment of the principle that from diversity, we can build an even stronger unity. And that in diversity, we can find the strength to prevail and advance."[2] I believe that may have been the first time I'd ever heard the word "diversity."

The official pronouncement was made, and the president and Ben's sister Elnora attached a fourth star to each of his shoulders. After the audience applause subsided, Ben stood at the podium, and the first word out of his mouth was a direct address to my father: "Judge."

Ben loved to call my dad "Judge." It was a respectful and loving acknowledgment of the connection they shared, that of a father and son.

Then Ben said, "When I came here this afternoon, I had no idea that the honors that have been leveled upon me in this room and to the United States Air Force, it is almost incomprehensible to me that I could have been here and not know anything about it."[3]

Many in the room laughed when he said he didn't know anything about being there. My parents and I looked at each other. We realized he was likely having an Alzheimer's episode and didn't remember where he was or what was happening. The speech was in front of him, yet he never looked at it. Instead he stated, "I didn't come here today to make a speech. I will not make a speech, but I will say to you, ladies and gentlemen, that all of us owe great . . . thanks to the United States Air Force for its contributions, which have not only been fought by them but by all Air Force people, and I thank you very much." The long pause after "great" and the somewhat unclear closing statement about the Air Force reinforced our belief that he was struggling, though no one else seemed to notice.

A couple of weeks prior, we'd accompanied him to Andrews Air Force Base for Thanksgiving dinner, as we did every year. Ben asked my sister, Sonja-Lisa, and me how long we'd been dating. My sister couldn't stop laughing, but I was confused and concerned. At one point, Ben and my sister danced together. Afterward, he came up to me and said, "She's a real keeper—I hope you two have a great family together." He was beginning to lose his recall of people, and we could tell his short-term memory wasn't fully functional either.

This was the beginning of the end, which made us fervently embrace every second with him.

After Ben's brief words, they took him aside for photos and to speak to some other invited guests. Clinton walked over and spoke to my family, shaking our hands and addressing us each by name, a gesture that really made us feel important.

This event happened at the same time I was working for Oscar Mayer, driving a Wienermobile across the country after I graduated from college. I hadn't received any other job offers. At my boss's suggestion, I'd brought President Clinton a present: a custom Wienermobile bank, which was tucked in a shoebox-sized container under my chair. I spent the event subconsciously mulling the perception of this gift and opted not to give it to him. Can you imagine the media response if I'd handed Clinton a Wienermobile, in the middle of his impeachment trial and the scandal of his affair with Monica Lewinsky? It would've made a mockery of the entire ceremony, and I didn't want to overshadow Ben. It would've been the equivalent of Will Smith smacking Chris Rock at the Academy Awards. Do *you* remember anything else from that night's show?

A fancy reception followed the ceremony. My dad, Ben, and John McCain had a long conversation. McCain put his hand on Ben's shoulder, looked him in the eye, and repeatedly said, "You deserve this." Ben did deserve it, yet when I reflect on the timing, particularly considering that it may have been politically expedient in some way, it was hard to separate the symbolism from the authenticity of the promotion. If you're not used to playing the game, it's challenging to navigate politics at that level. And maybe that was part of the reason Agatha had been so upset—there was some measure of tokenism involved that she didn't want to support, and she perceived the act as disingenuous. If they thought he'd earned the fourth star, then he deserved to be financially compensated for it. Why would you acknowledge the wrong but then make up for it in stature but not in compensation?

When we left the White House and returned to Ben and Agatha's apartment, Agatha was still in her pajamas—and still in her anger.

I'd been around Ben my entire life and was aware of his status in the military. But we conversed about mundane topics, keeping it light. He'd ask me, "How are you doing? How are your grades? What do you want to be when you grow up? How can I help you?" He bought me my first car, a white Honda Accord, gifted me his last set of golf clubs—my first set—as a hand-me-down, and gave me my first desktop computer.

Ben was incredibly humble. He never bragged about any of his achievements, and when he mentioned any of them, he credited those around him. And the man never complained. He'd eat rock-hard toast that had been sitting on a plate for hours and say, "Better than nothing."

That day, at the White House ceremony, his importance hit me for the first time. This warmhearted grandfather figure in my life, with his welcoming smile and infectious belly laugh, was a big deal. The president of the United States, the most important person in the entire country, had praised Ben's life in a way that was beyond anything I knew was possible.

On one hand, I was kind of sad that Ben was receiving this honor when he wouldn't be able to enjoy it for long. On the other hand, seeing him standing so tall at that podium, firmly shaking the president's hand and thanking him, made me happy and proud. It was the period at the end of the sentence.

Promises Made, Promises Kept

No matter how much information we take in, including about our own families, there's much we don't and may never know. Due to my lack of understanding of how the military worked, I had no idea what the refusal to compensate Ben meant.

Unfortunately, I don't have the income amounts for a three- or four-star general during Ben's years of service, but we can look at current Air Force pensions, which are typically based on years of experience and other criteria. Retired three-star generals' annual pension is about $189,792, versus the $215,104 of four-star generals, a $25,312 difference.[4] One can surmise the gap was similar back then. If you multiply these amounts by the number of years Ben would've been paid that money in retirement, thirty-four years, this totals an extra $860,608. When you factor in other benefits—medical coverage, a housing allowance, and so forth—this is why I refer to it as the "million-dollar star." It represents the financial compensation Ben should've received if he'd been promoted back in the 1960s, until his death in 2002. That income increase would have changed not only Ben and Agatha's lifestyle and retirement but also the financial freedom of the whole family.

Ben had promised his father he'd earn that prestigious fourth star. And he kept that promise, even in the face of his wife's anger. Yes, the million-dollar star represented financial gain that never manifested, yet it symbolized a deferred dream finally achieved too. The pinnacle of one's career, the culmination of one's life experiences. It's something you wear with pride—sometimes quite literally on your shoulder.

Unknown to me, that star would prove to be the key that would begin to unlock opportunities for future generations, including me. Without that fourth star, many of Ben's posthumous honors wouldn't have been possible, as you must have full general status to receive consideration for some honors. The most visible and prestigious monuments were reserved for four-star generals.

The occasion provided a context I'd yearned to know. My family operated in a sort of invisibility mindset. Silence is safety. Silence is security. Keep your mouth shut and do your job to the best of your abilities—let your talent and accomplishments speak for you.

Additionally, I realized that Ben had to have longevity. Only a few military members have been awarded a fourth star posthumously. However, just as America's first Black four-star general, Chappy James, died shortly after being promoted, Ben was so old when they promoted him that he was in a similar situation. Neither of these men ever got a chance to bask in this glory during their prime years or reap the full rewards of their status. It's a shame.

Ben had been promised many things over the years, but when these promises were broken, he looked on the bright side of things. My dad had adopted an "I'll believe it when I see it" attitude and tried not to get his hopes up. I witnessed this during my involvement with West Point, when they decided to construct and dedicate a barracks in Ben's honor.

NINE

Life after Death

What questions would you ask if you knew the end was near?

Arlington National Cemetery is the most famous burial place in the United States, and maybe in the world. Throughout the 639 acres, more than 400,000 tombstones mark graves as far as the eye can see.[1] Though the majority of those buried are American soldiers, their close relatives and a handful of other people are buried there—chaplains, doctors, and some civilians. It's the final resting place of Thurgood Marshall and former president John F. Kennedy, and every year on Memorial Day and Veterans Day, the current president places a wreath on the Tomb of the Unknown Soldier.

I was fascinated to discover that the land used for Arlington originally belonged to Robert E. Lee's father-in-law, and the federal government confiscated it at the beginning of the Civil War.[2] Lee is a constant in many strands of American history, whether we like it or not. I also learned that when Arlington first started burying soldiers, they weren't segregated. Blacks and whites were considered equals in death, even if they weren't considered equals in life. However, during the Civil War, they segregated the graves, a practice that continued until Executive Order 9981 was signed in 1948.[3] It's crazy to think of what we don't know or aren't taught. Segregating the dead? Really? That's deep.

Not far from the main entrance into the cemetery, atop a gently sloping hill, the grave of General Benjamin O. "Ollie" Davis Sr. is graced by the standard white tombstone. He's buried next to his second wife, Sadie Overton, to whom he was married for over forty years.

In the same section of the cemetery, only a few rows away, a much more prominent headstone stands out among the others. This headstone was hewn from black stone—a material remarkably comparable to the stunning granite used to construct Davis Barracks at West Point.

The tombstone bears four large stars and two names, of a husband and wife who were married sixty-five years. The wife, Agatha Davis, died on March 10, 2002, at the age of ninety-four. Her husband, General Benjamin O. Davis Jr., died a few months later, on July 4, 2002, at the age of eighty-nine.

Standing at Ben's tomb and glimpsing Ollie's close by is everything to me. It's a happy ending. It's the two Invisible Generals saluting each other for eternity, acknowledging the bond they shared, honoring one another's service, and saying to each other that *we* did it as a family. And even if no one else knows or sees it, we do. This eternal, invisible bond speaks louder than words.

An Audacious Request

After having lived an extraordinary life, Ben's final years were relatively ordinary. His Alzheimer's progressed rapidly after the fourth-star ceremony in 1998, and Agatha struggled with her own cognitive decline. Ben was the primary person who managed their household, but when he began to realize his mind was slipping, he wanted Agatha to take over the bulk of those duties. My mom, Sonja, remembers those years much better than I do, and she has shared a lot of stories with me.

She recalled once visiting them, and Agatha said Ben had wanted her to add some numbers on a calculator. Mom said she'd help her.

"Let's start with an example," Mom began. "What's two plus two?"

"What's plus?" Agatha replied. Ben was obviously unaware of how much Agatha had forgotten. Mom had a similar experience when trying to show Ben how to do something on his computer. He repeatedly insisted that he'd understood, yet couldn't replicate the simple task of saving a document.

Soon after, a major incident happened that forced the family to decide that Ben and Agatha needed someone close by to keep an eye on them. One day, they were out driving in the snow. Agatha navigated when Ben drove, but this time she couldn't remember the directions. They got hopelessly lost. My mom isn't sure how they found their way home. But one thing was for sure: that was likely their last attempt to drive anywhere.

Ben's younger sister, Elnora, felt she needed to be there, so she relocated from Los Angeles into an available apartment in Ben and Agatha's building. Even though Agatha was close to her sister-in-law, Elnora was overprotective of her older brother. Agatha would burn food on the stove, which Elnora would discover when she went to check on them and then reprimand Agatha. Initially, Agatha would let Elnora into the apartment, yet over time, she wouldn't even open the door for her. Agatha couldn't grasp that the situation they were in was hazardous—and the tensions were rising.

Eventually, it became clear that Ben needed to move into a living facility that could provide some medical care and supervision. Agatha heard about this plan and refused to be separated from Ben. "Not an option," she firmly said. They'd been one another's companion for so long that the symbiotic nature of their relationship made being apart inconceivable. When the time arrived, they were moved to a nursing home and placed in a room together.

Agatha's favorite words to hear were "I love you," especially from Ben. At that stage in their life, the only thing Agatha wanted to hear from him was those three words, which he'd said millions of times to her over the years: *I love you.* She'd ask him regularly, "Do you love me?" By then, he wore a hearing aid, and his mind had deteriorated to the point where he might not have known exactly where he was, or who Agatha was. He usually remained silent in response to her question or replied with a noncommittal "Uh-huh" or "Yes." She'd then say, "Why don't you love me?" or "Why can't you say that you love me?"

This would go on and on. And after a while, his lack of response enraged her. She'd yell at him while pushing and hitting him. This behavior alarmed the staff, who constantly called my parents to tell them what was happening and to ask if they wanted to come down and try to help keep the peace between Ben and Agatha.

Then everything came to a head. Ben's refusal to say "I love you" so angered Agatha that she went to the sink, filled a bucket with cold water, and dumped it over Ben's head. After this infraction, the nursing home informed my family that they had to move Agatha to a different room because her behavior was neither permitted, healthy, nor safe.

I feel it's important to share these family moments that can, and do, happen to any of us. I'm doing this in spite of knowing that Ben, as a proud and private person, probably wouldn't appreciate me sharing these intimate details of his final days with his beloved wife. In his autobiography, he had only praise for her. No person or relationship is perfect. Yet just as Agatha had opted not to include anything negative in the materials they donated to the Smithsonian, I believe that Ben chose to focus only on his wife's positive attributes.

He believed in highlighting the boundless potential of every human being, a true eternal optimist.

Agatha would pass away in 2002, a mere four months before Ben. I was overseas and thus was unable to attend her funeral. My parents later told me about the day. They went to the nursing home to get Ben ready. At least externally, he appeared emotionless. He said he thought Agatha was asleep somewhere. On the drive to Arlington National Cemetery, he kept asking where they were going, and though they explained to him, the reality didn't seem to sink in. When they arrived, something must have clicked, because he refused to get out of the car. They tried to persuade him, but when he wouldn't budge, my mom stayed in the car with him while my dad attended the graveside service.

After Agatha's funeral, my dad received a call from the nursing home that shook me to my core when I heard about it.

At night in the pitch dark, they would hear Ben crying out in his deep, baritone voice: "Agatha, I love you! Agatha! Agatha, I love you!" over and over and over again. The forlorn sound reverberated throughout the empty halls.

It's impossible to know what was going on in his mind. Was he having dreams about her? Did he hope that by stating his love for her that she'd reappear? That she'd embrace him and tell him everything was going to be all right?

Shortly thereafter, Ben was moved to Walter Reed National Military Medical Center in Bethesda, Maryland. At the end, he was moved to the area where President Eisenhower spent the last year of his life, in a special suite in the original hospital, overlooking the rose garden and fountain. This suite was where high-ranking military and some higher-level politicians stayed when they were admitted to the facility. In this symbolic, spacious corner room, Ben passed from this world, with the dignity and respect he deserved.

Beyond becoming a four-star general, Ben had wanted to die as an old man, peacefully in his sleep. And nearly four years after President Clinton and Elnora pinned those stars to the shoulders of his Air Force dress blues, his final mission was accomplished.

Of all days, Ben's last breath was taken on America's birthday. I tear up every time I think of it. He and his father had dedicated a combined eighty-four years of life to serve the country they loved in spite of its flaws. Neither of the Bens ever stopped believing in the promise of America and continually worked to make life better not only for Black Americans but for all Americans.

I cried when my parents called to tell me he'd passed away. One of the greatest minds in the history of the US military—one of the kindest, gentlest, most generous souls I'd been blessed to know—was gone.

Ben and Agatha were interred together at Arlington National Cemetery on July 17. The day was a fitting tribute, from the moment they carried his flag-draped casket into the room where the funeral service was held, to my dad's eulogy and the moment a red-tailed P-51 Mustang flew overhead at the graveside service.

At the funeral, my dad spoke on behalf of the family. My dad tended to focus on the positive and on the future, and he really didn't talk about the hardships anyone in our family had endured because of their race—a mindset he'd learned from Ben and internalized.

That day he delivered the speech of his life, talking about topics he rarely discussed—his relationship with Ben, our family, and racial issues—with a passion I'd never seen him exhibit. In that moment, he refused to be silent. His words are so powerful that I fear I could not do them justice on my own:

> *I have to say this—and I'm sure that although Ben would not approve of the style, he certainly would approve of the message. I have to make a rather blunt statement about American society*

in order to get across that message. And that statement is that there is a basic difference between whites and minorities in this country. And that distance, that difference, has existed ever since the beginning of American history and probably will continue for a number of years in the future. The difference is this: That in America, white men are born into the world with the presumption of respect, dignity, and honor. And that presumption continues with them until such time as they themselves, through their own disgraceful conduct, rebut that presumption. On the other hand, Black men, brown men, yellow men, red men, and women of all colors could not acquire those attributes through birth. They had to acquire them the hard way—they had to earn them. And Ben understood these rules of the American politics, and he was determined to overcome them. Not by demonstrating, not by denouncing, not by complaining, not by whining, but by succeeding. He was determined to succeed. And I think we all agree that he did.[4]

Although the white marble tombstones at Arlington National Cemetery are iconic, one of Ben's final requests was to have a memorable tombstone, carved from black stone.

When I visited Ben and Agatha's grave after the tombstone was installed, it was an impressive sight to see. He had asked for a plot that overlooked the white tombstones of the soldiers who had overlooked him in life. The tombstone and plot requests were very uncharacteristic for Ben. He was humble and never wanted anyone to feel bad. Selecting a black tombstone placed on a hill is reminiscent of the subtlety of painting "By Request" on the nose of his plane: quiet statements, both powerful in nature and legendary in history.

My father became the executor of their estate, and he and my mother cleaned out the apartment after Ben and Agatha were gone. During that process, they learned a shocking family secret.

Under the king-size mattress Agatha had shared with Ben for so many years, they found an envelope. Within that envelope was documentation confirming a medical procedure she'd had while Ben was stationed in Tuskegee. Agatha underwent a hysterectomy. No doctor was listed on the paper; the designated lines for that info were blank, either intentionally or due to someone's slipshod documentation.

This discovery rocked us. How could this have happened? Was it voluntary? We didn't know the answers to these questions, but we did know a couple of things. The first was that during Ben's time in the service, the military establishment had concerns over the continuation of the Davis legacy within the military. He was now a third-generation government/military family member, whose position and voice had gained more influence. He himself was too familiar with the system and thus too powerful.

Ben had confided to close family that the Army had told him he'd be dishonorably discharged if he'd had a son. He and his lineage had become a threat to the status quo. This revelation both shocked and disgusted me. But it was also a reminder that, in some ways, the military "owned" Ben. Sure, he could've left at any moment, but a dishonorable discharge would've been a permanent stain on his record, limited his opportunities in the private sector, and prohibited him from being recognized for everything he accomplished.

Another thing we knew was that the American government has a long, sordid history of using women, Blacks, veterans, and other vulnerable groups as unwitting participants in medical procedures. Ben's time in Tuskegee—including while the Airmen trained there—overlapped with the "USPHS Syphilis Study at Tuskegee," more commonly known as the Tuskegee Experiment, conducted from 1932 to 1972.[5] About 600 men were enrolled in the program without their knowledge or consent; 399 of them had syphilis. The men who had syphilis received no treatment. More than 100 of them died from this treatable disease, and some infected their spouses, who then passed the disease to their babies.[6]

Though the Tuskegee Experiment was specific to men, women were subjected to equally inhumane treatment by the medical establishment. In the twentieth century alone, some women were forcibly sterilized—almost 70,000, a staggering number. Most of the victims were poor, nonwhite, and/or those considered mentally deficient, and a Supreme Court ruling had established that states could force sterilization on people considered "undesirable" to procreate.[7] Sometimes women were institutionalized and then sterilized; others were sterilized when they were admitted to the hospital for routine procedures.

Had Agatha been victimized by this cruel practice? Regardless, why didn't she ever tell anyone in the family? Was the shame too great?

The letter answered the question of why Ben never had a biological son, yet it presented many more questions.

The biggest question of all was what Ben knew. Would Agatha have told him? Or would it have been too much for him to handle? I can assure you: he *always* wanted a son, so there was no way he would have cosigned it. And I doubt that with Agatha's tremendous love for him and their intense connection, she would have done it voluntarily, unless she was in a desperate situation. However, the medical field wasn't as advanced then as it is now, and hysterectomies were considered the "cure" for all sorts of diseases afflicting women.

For better or worse, my dad's fate was tied to that piece of paper, which was discarded with all the other items we didn't keep. It was a painful reminder of yet another family story lost to time and the shadows of secrecy.

I wish I'd spent more time with Ben in his final years. That's the funny thing about time, especially when you're young—you always think you have more of it coming. Ben might not have even known who I was when I went to see him in his final years, but I knew who *he* was. I had no way to foresee the ways his larger-than-life existence would one day influence every aspect of my own life, or that after he was gone, I'd end up being his representative, working to get him the credit he'd deserved while he was still here.

TEN

Monumental

Every generation is a continuation of the previous generation's evolution.

When Ben graduated from West Point in 1936, it had been almost fifty years since a Black cadet had done so. Charles Young, Ben Sr.'s mentor who'd been the first Black colonel in the US military, graduated in 1889. Between the time these two men attended West Point, one group more than any other, the United Daughters of the Confederacy, made a concerted effort to reinsert Confederate war figures on campus, while downplaying these men's participation in the Civil War. Robert E. Lee's name and visage was frequently deployed, and in 1930, the group donated a portrait of the general to hang in the mess hall. In 1931, they began to sponsor an annual math award named in honor of Lee.[1]

Interestingly, Lee himself was against such displays. He "opposed building public memorials to the rebellion, saying they would just keep open the war's many wounds."[2] However, after he died, he was adopted as a symbol of Southern pride and the white supremacist cause, much like the Confederate flag. As with pretty much everything in history, Lee's life and legacy are complicated.

The most egregious Lee display marked the culmination of the revisionists' attempts to rehabilitate his image and legacy, when West Point celebrated the one hundredth anniversary of Lee's becoming superintendent as well as the one hundred fiftieth anniversary of the institute's founding: "On January 19, 1952, a massive portrait of Robert E. Lee—in full Confederate gray uniform, with a slave guiding his horse behind him—was donated to the West Point library."[3]

Lee's ever-heightened visibility stood in stark contrast to West Point's most invisible alum.

Understandably, Ben only visited West Point a couple of times after graduation day. He returned to campus for the second and final time in 1987, the year after his fiftieth reunion, to spend a week researching for his autobiography. Prior to that, he'd received a letter from the superintendent, indicating that West Point wanted to display Ben's portrait and label it "America's First Negro Air Force General," but Ben told them to eliminate "Negro." He wanted to be known merely as a "general" or a "great American general," not as a "Negro general." He believed that when you put "Negro" or "African American" in front of something, you demeaned that person by labeling them. He wanted his portrait to be displayed in an equal way to all the other ones, without qualifiers. As he wrote in his autobiography:

> *Some black Americans, Rev. Jesse Jackson among them, have suggested that blacks be called African Americans. As President Theodore Roosevelt remarked some decades ago, Americanism is diminished by this kind of divisive nomenclature. We are all simply Americans. Differentiations created by extensive statistical studies for no obvious purpose except to prove the superiority of a particular group are not helpful, and surely the unnecessary labeling of people by race, religion, or ethnicity does nothing to bring the many diverse groups of American*

society together. . . . I do not find it complimentary to me or to the nation to be called "the first black West Point graduate in this century."[4]

Ben knew that many disagreed with his perspective on the matter, and some even found it controversial. Nonetheless, he refused to capitulate. He even threatened to cancel the publication of his autobiography in 1991, when the Smithsonian Institution Press wanted the subtitle to include "African American." He told them it must be "American." Period. And he prevailed.

In my family, West Point was spoken of almost like a bogeyman. Of course, we knew that Ben would not have been able to accomplish everything he did in his career without having attended West Point. The institution's prestige was an undeniable asset for anyone in the military. Yet because of the way he'd been treated there, the relationship had become strictly transactional.

Before I learned about my family's history, I had zero awareness of or concern for preserving the legacy of those who came before me. Sometimes when you are young, that is what you do. But the process of unpacking my family story slowed me down. In the short term, it was like cleaning out your house, and instead of shoving the boxes in the attic and moving on, you stop and look through every single box, taking stock of each item. In the long term, this experience changed the direction of my life.

After the *Red Tails* screening in 2011, when I flew from Dallas to New York City, I couldn't wait to rant to my dad. Something about Ben's being erased from his own story had filled me with a righteous anger. When my plane landed, and I set a Google Alert for "Benjamin O. Davis," I didn't add *Sr.* or *Jr.* because I wanted to receive notifications for anything that came up for either of them. Then I would contact anyone who posted anything about either Ollie or Ben, to let them know that even though Ben was deceased, his family was alive and well, and we wanted to be involved in the shaping and presentation of anything surrounding his name, likeness, story, or legacy.

I received an email confirmation that the Google Alert was set, and I eagerly waited.

And waited.

Years passed—2012 . . . 2013 . . . 2014 . . . I began to wonder if the alert was even working.

Finally, in November 2015, the alert went off. An article in a small upstate New York newspaper reported that West Point was considering naming a new barracks after Ben.

Right away, I contacted the US Military Academy, the Army, and the Department of Defense, asking each to let me visit West Point and talk to them about the building dedication. Though I had a vested interest in advocating for the dedication, my primary goal was to have an opportunity to tell them the true Ben Davis Jr. story, the one I had seen up close and had unearthed from talks with my dad. I wasn't sure what, if anything, they knew beyond what was published in print media and online.

When I told my dad I was planning to visit West Point to discuss Ben, he was surprised and a bit taken aback. He said, "Doug, you have a full-time job. Why are you wasting your time with all this foolishness?" I don't think he realized how much Ben's story had consumed me since that *Red Tails* screening four years prior.

Part of my role as a diversity officer involves ensuring representation, so I appreciated how critical that piece was. For me, it was about the respect Ben had earned and deserved. He'd wanted nothing more than to live to be an old man, so he could serve his nation and fellow Americans as long as possible. But the only way he could do that was by intentionally choosing to remain invisible—to not draw attention to himself and thus not be perceived as a threat.

Most important, this was about family. A huge part of the reason we know Martin Luther King Jr., Malcolm X, and other now-prominent Blacks is through the advocacy of their children or family. I wanted to do the same for Ben. If the military wasn't going to compensate Ben or my family with money, perhaps it could compensate us through righting some wrongs of the past, by making Ben visible and allowing this recognition to cement his legacy in the current and future generations. I made it my personal mission to become his publicist, taking on the mantle to ensure that in death, his voice and vision are louder than they ever could have been in life. I wanted to ensure that he and his accomplishments were visible.

One month after the alert went off, December 2015, I was scheduled to meet with a handful of people at West Point's office of diversity. The fact that they had created an office of diversity, equity, and inclusion was a good sign. It happened to be Ben's birthday week (he was born on the eighteenth), so I arrived with a box of celebratory cupcakes to mark the honor. If the meeting went well, I wanted us to sing "Happy Birthday" to Ben.

As I drove to the campus with a friend who'd joined me to document the day's events, my mind began to wander. This was the first official meeting between my family and West Point. I had no idea what would happen. Would I receive an icy reception? Would the conversation be combative? Would it be joyful?

We pulled up to the building and were greeted by a parking spot bearing a sign with my name on it that said "General Davis Family." I was shocked. I wasn't sure what to expect, but it wasn't the private meeting I'd anticipated.

Upstate New York in mid-December is brutally cold. That day, I think it may have been sleeting. In spite of the nasty weather, the campus was much more beautiful and impressive than I would have thought. The massive, historic buildings stood sentry, as if the place was a time capsule left completely untouched by the modern world.

Several people greeted us in the meeting room, including the acting chief diversity officer and a representative of the Association of Graduates. As we conversed around a long table, they told me some background about the project. West Point had been working with the Army Corps of Engineers to construct what would likely be the last brand-new barracks—the largest, tallest barracks in the campus's history.

They'd narrowed down the list of potential honorees to three cadets who had risen to the rank of general: William Westmoreland, one of Ben's classmates who'd commanded troops in World War II, the Korean War, and the Vietnam War; (Herbert) Norman Schwarzkopf Jr., who was most famous for his leadership during the Gulf War in the 1990s; and Benjamin Davis Jr. Whereas Westmoreland and Schwarzkopf had been active-duty four-star generals, Ben had only reached the rank of three-star general while he'd been on active duty and had received his fourth star in retirement. Of the three candidates, Ben was the least qualified—at least on paper. Even so, Ben's name had taken the lead because West Point wanted to highlight diversity in the naming of the building.

Diversity. They were speaking my language. Working as a chief diversity officer for the past few years had prepared me for this moment.

I began to shed the nerves I'd felt when I walked into that room. Now I knew I could share a fresh perspective and weave a narrative that would not only give a voice to what Ben endured at West Point but also convey the pride my family had in his achievements and our commitment to preserving his legacy.

By the time I finished telling the story—about the Invisible Generals, a father and son, about my dad, about me—there wasn't a dry eye in the room. Among the group were cadets of color, who had worked long and hard to attain their status, and I have zero doubt Ben's story resonated with them. It became clear they would do everything in their power to ensure that Ben's name was on the barracks.

Before I left, we did get around to that "Happy Birthday" for Ben and lightened the mood over the cupcakes I'd packed with me.

The meeting officially ended and they thanked me for coming, yet I had one more bomb to drop on them: I'd noticed that West Point, now contemplating naming a building in Ben's honor, had failed to even list him in the online directory of graduates. Not surprisingly, they didn't believe me, so I pulled up the website to show them. Lo and behold, the West Point graduate directory listed every single cadet except Benjamin O. Davis Jr. If nothing else, I'm sure that this revelation reinforced everything I'd just told them, and lent credence to the idea that his exclusion had been the result of a conscious effort to systematically delete him from West Point.

The West Point barracks naming was both the biggest opportunity and biggest challenge. Before then, nothing on campus had been named for any of West Point's Black graduates. Fortunately, everything was coming together in the right place at the right time. I was living in New York City and had a relatively easy drive to the campus, which meant I could visit often. Another advantage was that West Point's office of diversity, led by Colonel Rod Doyle, was a tremendous advocate for the cause, as was Archie Elam of the Association of Graduates. The folks at West Point knew I was well-connected, especially with the media, so I think they probably wanted to ensure their dealings with the family didn't result in any bad press or negative perceptions of their efforts.

At the beginning of almost every meeting with the team, I'd tell them that if Ben were there, he'd express his appreciation for all they'd done—but I wasn't Ben. I was dealing with the military industrial complex, and as far as I was concerned, all that kumbaya stuff wasn't going to cut it.

As expected, some obstacles emerged as the process evolved. The first voices of opposition came from alumni, who wrote in saying that if a building was going to be named after Davis, it should be an older barracks and for the new barracks to be named after George Washington or Robert E. Lee. When that didn't fly, alumni suggested a generic name like the "Barracks of the American Soldier." But then the current cadets got involved and initiated a grassroots campaign, encouraging their fellow cadets to sign a petition indicating that they wanted Ben's name on the building, a document they presented to the superintendent.

During the construction process, the US Army Corps of Engineers, the group responsible for building the structure, had invited me to observe their progress. At one point, they were concerned that one of the chapels was going to slide down the hill on which it sat because they'd blasted and drilled too close to the structure when clearing the land for the barracks. Even those who were excited and eager to have the new building named after Ben were less enthusiastic about the almost 24/7 noise disrupting their studies, lifestyle, and sleep.

If that wasn't enough, the government froze military funding after President Trump took office in January 2017. Davis Barracks would be the largest, most expensive barracks ever erected on the West Point campus, and the new administration paused construction until they could evaluate all existing spending efforts. This moved the projected completion date from spring to summer 2017, pushing back the ribbon-cutting until the funding issue was resolved.

During the years I was engaging with West Point about the barracks, a national debate erupted around Confederate monuments. Politicians initiated legislation to either retain or remove these, and one particular statue—of former Confederate Army commander Robert E. Lee in Charlottesville, Virginia—became a lightning rod. Tensions in Charlottesville reached a fever pitch over the planned removal of the Lee statue from the center of the city. On August 12, white supremacists organized a "Unite the Right" rally to express their outrage. Violence erupted, and a member of the white supremacist group drove into a crowd of counterprotesters, killing one woman and injuring dozens. The powder keg of racial animosity had exploded, hastening the removal of Confederate monuments and statues across the country, whether by government or fed-up activists.

On August 18, less than one week after the Charlottesville tragedy, hundreds of people, including high-ranking military officials from all branches, gathered at West Point to attend the ribbon-cutting ceremony. The General Benjamin O. Davis Jr. Barracks officially opened as a cadet residence hall, prominently placed in the middle of campus. This massive, imposing structure was designed to match the historic Gothic Revival buildings already adorning the campus, and almost two hundred thousand square feet of beautiful black granite was handcrafted for various parts of the building.[5]

A lot of energy had built up to and in that moment. Standing on that podium, my dad and I were equally overwhelmed, as we represented two different perspectives. For me, this was about the cultural and historical significance of the day. On some level, I felt the rage of Charlottesville inside me, an accumulation of many generations of indignities and oppression and silencing. However, I kept this emotion inside and chose to view the moment through the lens of the future. I thought about the process leading to that moment and the people currently at the Point. I also thought about the Black cadets who needed to be encouraged and inspired by this, and what all the cadets could learn about and from Ben.

For my dad, this was personal. It was about the man he'd known his whole life, a father figure who'd raised him and whom he'd seen struggle against forces that'd tried to keep him down. He was viewing the moment through the lens of the past and his lived experience. He kept saying it was too bad Ben wasn't there to see the barracks, and how Ben wouldn't have believed it.

The presence of media there to cover the event showed my dad that people who weren't directly involved with Ben or West Point were interested in what was happening, and I was gratified to see my dad's cynicism of my "chasing the ghosts of the family's past" further transforming into excitement. Over and over again, he thanked me. He couldn't stop smiling, his whole being infused with joy and pride. I felt that the moment belonged to my dad, and in many ways, this whole pursuit was a gift from me to him.

I'd insisted to the Army Corps of Engineers that all the workers who'd contributed to the barracks be invited to the ceremony, and quite a few of them attended. One of the guys who had worked on the construction led me about two hundred feet from the entrance and showed me something he'd secretly etched on a stone, a version of Psalm 118:22 from the Bible: "The stone which the builders rejected has become the cornerstone."[6] He told me that when he read about Ben's experiences and accomplishments, he was so inspired that he created a cornerstone. I learned that many of his fellow builders had likewise been inspired, and I was again reminded of the power of Ben's story and further motivated to keep moving the family narrative forward.

The last brick was inserted that day in August 2017, and it contained a time capsule full of documents, photographs, and mementos. It's scheduled to be opened one hundred years after the final brick was put in place, and in my wildest imaginings, I can't envision what

West Point—what the entire world—will look like then. The words spoken by Colonel David A. Caldwell, who commanded the Corps of Engineers, resonate deeply: "It's not just a building on campus, this is a monument, it's one of those forever buildings . . . it will be standing long past any of us or our children or our children's children."[7]

After the event, someone from the Air Force, whose name I can't recall, came up to me and said, "You know, this is a great dedication to General Davis, but he was really an Air Force guy. We need to make sure we do something about that."

I then started to meet with Air Force officials, who wanted to dedicate the US Air Force Academy airfield in Ben's honor. The airfield, which is located at the center of the USAFA campus, is one of the top five busiest airfields in the United States, based on flight traffic.

At my first USAFA visit, they seated me in the superintendent's conference room. Though I wasn't meeting the superintendent that day, I did speak with a thirty-plus-year employee who worked in the main office, who was known as the unofficial boss. While we were talking, I noticed many portraits of Air Force generals displayed all over the walls, but none of them were Black—and none of them were Ben. I told Ms. Gail Colvin that I'd love for them to hang Ben's portrait in the superintendent's conference room. We then dove into the details of the airfield dedication.

Two years later, on November 1, 2019, we attended the airfield naming ceremony. At that event, Air Force chief of staff, General David L. Goldfein, said, "Some have had to bear a heavier burden than others to teach us all what right looks like. Today, we celebrate one of these men."[8] And my dad made a remark that sounded like Ben himself: "Thank you, Air Force Academy, for bestowing this high honor on General Benjamin O. Davis, Jr.—American, full stop."[9]

At the dedication ceremony, I again brought up the idea of Ben's portrait in the superintendent's conference room and had my dad mention it to people too. I much prefer that storytelling style than sending strongly worded emails, because I think people are moved and motivated by relationships, connection, and the emotions of a good story.

About a year later, I heard that they'd added Ben's portrait to the wall. Although I have yet to see it in person, this news brought joy to my heart, knowing that he'd been invisible in that space for so long. I want him to be integrated into the look and feel of different walls of honor. There doesn't need to be a big announcement—just subtle changes over time, nudges and inches to get to the point where people know his face and name and want to learn more about him and his contributions.

Interestingly, the Air Force has even chosen to honor Ollie. In August 2022, they dedicated Quarters 64 at F. E. Warren Air Force Base to him, because he lived there in 1912, when he was a Buffalo Soldier. The ceremony involved Buffalo Soldier reenactors and the unveiling of a rock memorializing Ollie, placed in front of the building. I wish I'd known about it and attended. The former curator of the base's heritage museum explained how important the Buffalo Soldiers—including Ollie and Charles Young—had been to Wyoming: "It was the Buffalo Soldiers that were able to calm the communities down and to get peace started. . . . So, the memory of the Buffalo Soldiers is extremely significant to the heritage of our state."[10]

In the coming years, the Air Force plans to upgrade the entrance to the airfield, with signage that more prominently shows Ben's name and a series of pieces that tell the Tuskegee Airmen story. Ben would've insisted on sharing the honor with his fellow Airmen, so I know their inclusion would've made him very happy.

That the centerpiece of both the Army and Air Force academies—one Ben had attended and one he had helped shape—bear his name is an achievement I couldn't have dreamed of. At both institutions, he's now visible in ways he never was in his lifetime.

After several years of racially charged events, culminating with the death of George Floyd in 2020, instead of taking for granted that something was named after a historical figure, people began asking *why* it was named after that person. In 2021, Congress established a naming commission to evaluate Army bases, military academies, and Department of Defense fixtures named after Confederate soldiers or that represented the Confederate cause. The commission identified nine Army bases and other fixtures for renaming and removal. Although Ben's name didn't make the final cut (I can't deny that Fort Benning being renamed for him would've been supreme poetic justice), Ollie was on the short list. The group submitted their final report to Congress in September 2022.

As for the removal of the Confederate references on existing bases, I knew this process was underway, yet I was still surprised on the morning of December 22, 2022, when Archie Elam texted me a photo of a brick wall exposed by a removed piece of artwork. This installation, which depicted about a century of American history, had made headlines in August that year, because its various elements included a hooded Ku Klux Klan member. He informed me that the removal of Confederate items had begun on Sunday, December 18—what would've been Ben's one hundred tenth birthday.

ELEVEN

The Brand Equity of Fairness

Some of the greatest lessons
for advancing forward are found
in looking back.

I'm not inherently a fan of the word "brand," mainly because of its connection to the enslavement of human beings and livestock being physically branded with hot irons to mark them as another person's property. Since then, the word has evolved, but it's ubiquitous. For someone who has spent a career in marketing and advertising, I can't deny the term's utility. When you say "brand," even if someone can't define the word, they can rattle off dozens of brands. And after this experience, I definitely consider the Tuskegee Airmen and the Red Tails two powerful brands.

Ben's proudest moments were the times and experiences with the Tuskegee Airmen. His legacy and their legacy are forever entwined. One thing that amazes me is that the Tuskegee Airmen seem to be showing up more and more these days. I don't know if that's because my awareness is heightened due to my interest in the subject, or if they've genuinely had a pervasive presence in recent years.

To my surprise, in 2020, when I tuned in to watch Super Bowl LIV, I saw Colonel Charles McGee presenting the official coin for the pregame coin toss. He was one hundred years old and one of the few surviving original Airmen. Three other centenarian World War II vets joined him, and in the sea of people on the football field, Colonel McGee's bright red blazer, worn to honor the Red Tails, stood out. Two days later, McGee was honored at President Donald Trump's State of the Union address and promoted in his retirement from colonel to brigadier general—a rank he'd earned in war that was denied to him because of his race. He received much applause and a standing ovation, and you could tell he was emotional and proud. If he was anything like Ben—and I suspect he was—perhaps the attention embarrassed him a bit. The Tuskegee Airmen typically weren't the kind to brag about themselves. Like their steadfast commander, they let their performance speak for them.

Based on Ben's experience of being promoted to a four-star general without being paid, I wondered whether McGee or his family received any compensation. I looked up the piece of legislation online. The first section stated that the president had been authorized to grant McGee the honorary promotion. Sure enough, the second section said the following: "No person is entitled to any bonus, gratuity, pay, or allowance by reason of section 1."[1]

Charles Young, Ollie's mentor, was posthumously promoted to brigadier general in 2022, one hundred years after his death. He was yet another Black military officer who deserved to be promoted to general during his long and storied career, but was forced into retirement. At the ceremony, held at West Point, Young's great-niece spoke on the family's behalf. She'd been working behind the scenes for almost fifty years to secure his promotion, and "she expressed that the Young family does not desert their goals and dreams, no matter how long it may take."[2] And once again, when I looked up the bill online, I read that the family wouldn't receive any money.

What happened to McGee and Young was exactly what happened to Ben. He'd served in the military over thirty years, and a postretirement promotion decades later was granted due to the injustices of his time, but the reward would be received with the denial of any financial benefit. Acknowledged, but not compensated. Was this fair? Is this just the way it is? I wasn't sure I could accept that. Part of my mindset around these issues might stem from my experience in the business world, where people are expected to and do receive compensation for their skills and contributions.

There's much to be said about the value of humility, of gratitude for honors received. Yet considering the billions of dollars wrapped up in the US defense budget and the military's commitment to marketing themselves at events like the Super Bowl—and this doesn't even factor in the seemingly limitless budgets in corporate America—it blows my mind that none of the Tuskegee Airmen or other great American men and their families has ever been able to cash in on their fame.

No Royalty for Loyalty

In 2021, the United States Mint announced the final coin being released in their "America the Beautiful Quarters" program. For ten years, the Mint has been issuing a quarter for each state or territory, with the reverse side of the coins depicting a national historic park or site from that area. The quarter for Alabama features the Tuskegee Airmen, with an image of a pilot suiting up, an airfield building, and two of the red-tailed planes in flight. Most poignant to me are the words inscribed on the tail: "they fought two wars," referring to the "double victory" concept of fighting fascism abroad and racism at home.[3] When I reflect on the way the Airmen were never able to capitalize on their renown, yet are used at every opportunity that can benefit someone else, I must ask: Are they still fighting a war at home?

After World War II, many Airmen continued to wage war against racism and segregation on the home front. Those who stayed in the military were often assigned duties they felt didn't align with their experience and status as decorated combat veterans. Others fared worse in the private sector, where opportunities were limited because racial discrimination in hiring was still legal in the US until 1964.

Another reason they never could capitalize on their success was because most people had never heard of them. Other than the Black newspapers, no one really covered their progress and feats in the war. Ben was once featured in a video for the Army Air Force News, where he told the reporter that though the interviewer might not be aware of it, he'd commanded an all-Black fighter pilot squadron in World War II.[4] Ben had been invisible for his entire military career, even as a multi-star general. Of course, those he commanded would be equally anonymous. Ben shared that they didn't even know until after the war that they were called the Red Tails by so many, until some of the Airmen were honored at a reception with a red carpet at an event, and that's how they were announced.

I feel there's a bit of a false perception about the popularity of the group during and following World War II. If you ever see a picture of the Airmen at various events, wearing their red jackets, that doesn't reflect the reality of their lived experience, or that the story of the Black fighter pilots who'd helped the Allied forces achieve victory was buried for years.

The Tuskegee Airmen didn't exist as a formal organization until 1970, twenty-five years after World War II. Charles McGee founded a nonprofit called the Tuskegee Airmen, Inc. (TAI), to raise money, awareness, and support for the surviving Airmen, along with their families. As another Airman, George Hardy, shared, "At that time, a lot of people in this country, even in the black community, didn't know that black people had flown in World War II."[5] The group took the name from the Tuskegee Institute and the Red Tail moniker, and began wearing the red blazers as a way to honor their history in a modern, classic style.

Though they might not have said they were crafting a brand, that was the outcome. I'd argue that they're the most recognizable fighter squadron from World War II. Other than military historians or fans, you'd be hard-pressed to find someone who could name another World War II squadron.

If people want to use the "Tuskegee Airmen, Inc." name, they pay a fee or make a donation to TAI. Yet as a nonprofit, the entire staff is volunteers, so pursuing licensing agreements and creating other types of lucrative relationships isn't the group's core mission.

The thing is, in spite of the Tuskegee Airmen brand being both powerful and valuable, neither the Airmen nor the organization "own" exclusive rights to many of the ventures that profit off their brand. They have the brand, but not the *brand equity* that translates into being able to leverage the "product" into financial gain.

Something important I want to be clear about is that I'm not pointing the finger. I'm here to criticize the system, not the players. But I believe that the Airmen do not get paid their worth. As of right now, almost anything created for or about the Tuskegee Airmen or about Ben is in the public domain. Thus, no one has ownership, and no one receives any money anytime something related to their story is sold. TV shows, movies, and documentaries can feature the Airmen, their likenesses, and even their names, without them getting paid.

Ben was a general and therefore earned more money, but the majority of the Airmen he commanded couldn't garner the same for themselves. Given what they sacrificed and endured to defend their nation, and the amazing feats they accomplished, it's unfair that their families have struggled so much financially.

Additionally, there seems to be tension between what the Airmen consider acceptable and what their families believe is just. I've heard from several people connected to the Airmen, and because of what transpired after World War II, many of the families have to rely on some form of public assistance. Sadly, this lack of visibility and lack of support, financial or otherwise, is the reality for many US veterans, regardless of race.

Even though the Tuskegee Airmen story itself isn't new, with the cultural tide shifting to greater awareness of and interest in contributions of nonwhite Americans, people see an opportunity to use the Tuskegee Airmen story as a free marketing campaign to generate goodwill.

Several years ago, I had the pleasure of meeting with Brigadier General Leon Johnson (retired), who was TAI's president at the time. We connected like we'd known each other for years, and his collection of Tuskegee Airmen memorabilia and collectibles is *incredible.* In our conversations, he frequently expressed his appreciation for George Lucas bringing the Tuskegee Airmen story to the big screen, especially since the project was a labor of love for Lucas.

During Lucas's media tour for the movie, he revealed that he'd been working on the project for twenty-three years (what?!), and he had to finance much of it himself. When it was finished and ready to go, studios still passed on it. Lucas attributed this in part to the fact that it was the first all-Black action movie, and the studios said they didn't know how to market it. While some of the original Airmen served as consultants, as the project dragged on, several of the Airmen passed away. Lucas felt that pressure, to try to wrap things up before they and their first-person accounts of the events were gone forever.

In an interview on *The Daily Show with Jon Stewart*, Lucas likened trying to write the script to his experience writing *Star Wars*: "I wrote the first script, I tried to get it to work. It was way too big. The story is too fantastic and wonderful to cram into two hours."[6] Toward the end of the segment, Stewart asked how the Airmen reacted to the movie, and Lucas replied, "They're overwhelmed because they've been waiting so long to be recognized. Most people don't know who they are. And they are really true heroes. I worked with all these guys—they're great, great guys. And just to be recognized. Just to be able to say, 'Tuskegee Airmen,' and say, 'Oh, I know who they are.'"[7]

Again, they were grateful for the recognition. I'm appreciative as well, but to me, the biggest tragedy is that millions of dollars were spent making this movie and collectively paying the actors portraying the Airmen, yet the Airmen themselves—the men who fought in the wars—didn't get an equal share of the financial opportunities the movie provided. It's the same situation with other Black stories that became bestselling books that also became movies, like *Devotion* and *The Immortal Life of Henrietta Lacks*.

But as Oprah said to me at the screening after-party, "Douglas, this is Hollywood." It's not anyone's fault per se—it's the flawed and unjust system.

It's probably unheard of that a for-profit movie studio would share proceeds and give back-end points to a nonprofit organization representing the people whose story is being told or to the people themselves. Even so, I do know General Leon Johnson was in contact with Lucasfilm, having conversations about the possibility of the trademarks or some kind of ongoing royalty being given to TAI for the movie. Leon represented the Airmen and the Red Tails brand at the highest level. He was the maestro of the organization. Yet even with his exceptional leadership, when Disney acquired Lucasfilm the conversations stopped, and the momentum fizzled out.

You can see both sides of this situation. On one side, you could say that the filmmakers don't owe anybody anything. They owned the trademark for the movie and didn't use any of the real people's names, so they're not legally mandated to pay anything. It can be said that getting the story out there *is* the opportunity, which may have greater value in the long term than a one-time payment. On the other side, you could say that if we live in a world where companies are striving to be more socially conscious and responsible, to improve their brand reputation and increase their revenue, wouldn't studios want to be equitable and dole out the money, to be fair and for the positive publicity?

The moral high ground is the most expensive real estate in the world. All of this goes back to my purpose and passion of being a diversity officer. I want to be the person in the room advocating for those who don't have a voice or power, and address both sides in a way that provides space and grace and leads to understanding and forward progress.

And companies aren't the only entities leveraging the Tuskegee Airmen brand. The military does, too, and this has been showing up more and more, particularly in recruitment efforts. As demands for diversity become the social norm in the United States, shining a light on the Airmen and Ben is a way to build a bridge between the past and the future. It's just unfortunate that these opportunities to be showcased don't come with an honorarium.

The military typically doesn't pay anything to use the Tuskegee Airmen, the Red Tails, or Ben's name. But I feel the Airmen can and should be an exception. The Tuskegee Airmen were different—they stood for something different, and their accomplishments are timeless. Equity is not equality, and their sacrifices were disproportionate and unfair. Also, they're tremendous assets, not liabilities.

In all the decades Ben worked for the military and government, all his ideas and creations became their property. He could never patent or trademark anything or receive any kind of lifelong residual payment or royalties. Even when he did receive royalties, such as for his autobiography published over thirty years ago, the total royalty amount received for ten years of book sales was $955.60; and the advance paid for his entire life story was $1,500. The Smithsonian Institution Scholarly Press has never modernized the book or sold it in newer formats, like ebook or audiobook, and they only print a handful of copies per year. In fact, I offered to buy the rights back myself, so his estate could own it. The email reply I received made it clear that selling the rights wasn't an option and never would be. Their reaction led me to believe that they're more interested in controlling Ben's story than they are in sharing it.

While attending the Davis Airfield renaming event in 2019, I got word from Leon Johnson that Google wanted to connect with me. They were creating a commercial for Black History Month 2020, featuring the most-searched Black Americans in various popular categories. It was scheduled to air during Super Bowl LIV, ensuring that it would be visible to millions of people. Once we connected, I was told that Ben was the most searched Airman on Google. This was news to me. I'd been promoting Ben as part of the Invisible Generals story for a few years, and I'd felt like I was working in a vacuum. While people had heard of the Tuskegee Airmen, outside of World War II history buffs or the military, no one seemed to have heard of Ben.

She told me that they had been in contact with Getty Images, which owned two images of Ben that Google wanted to use. Getty wanted to charge around $100,000 to license the files, but Google was trying to negotiate the price down. In the meantime, she was contacting families of those included in the commercial, to see if they'd be willing to supply images or video. I asked her how much the families were being compensated for providing this material, and she told me that the "favored nations" blanket licensing agreement more or less meant that the name and likeness would be donated, and the "payment" for everyone would be the visibility.

I had a choice to make. Was I going to practice what I preach? I'd spent years saying that Ben needed to be financially compensated in these types of situations. People who are historically oppressed are often compensated with "opportunity" in lieu of money. This was the million-dollar star all over again.

The Google team had explained that the spec commercial was running long, so some people who were featured would need to be cut. If I forced the money issue, I had to consider whether they would exclude Ben.

I decided to press on in the way I'd had the greatest success: I told her the family story and emphasized that Ben did all these amazing things, yet was never compensated fairly for them. The US military media department had followed Ben and the other Airmen for years, incessantly photographing and filming them to try to capture their failures. I didn't find it fair or equitable that Getty Images had bought a huge lot of World War II photos and videos from the US military and proceeded to go around selling the rights to use them—all without having to pay the families. The US military got paid, yet the subjects or their family didn't even get a courtesy call. I understand how this system works, but that doesn't mean I agree with it.

In the end, they wanted to use Ben in two commercials, one for a year-end review spot and the other for Black History Month.[8] Google agreed that they would compensate us $20,000 total for both commercials.

When I shared this exciting news with my dad, he was suspicious and interrogated me like he might have a witness back in his courtroom days. *Ben's dead—why are people still searching for him? Who's searching for him? Why would anyone pay $10,000 for a three-second clip for a commercial? Are you sure this isn't some kind of scam?*

Dad remained reticent, but seeing is indeed believing. When the two checks from Google arrived, I persuaded him to travel to Washington, DC, so we could meet with the attorney and reopen Ben's estate, appointing me executor. By then, the estate had long since been liquidated and closed. It had a zero balance—no assets, no money, nothing. I told my dad I wanted to reopen the estate in case other offers came in, because I believed I could sell Ben's name and likeness and start accumulating some funds.

"We can bring Ben's legacy back to life—and make him visible," I said.

He nodded and said, "You already have, son. Even if nothing else ever happens moving forward, he would've been so proud." For the first time, my dad was hopeful that we, as a family, could find ways to continue honoring Ben. And his joy became my joy.

On the train ride down to DC from Connecticut, we had a chance to bond. I wondered if this trip was in some ways similar to those Ben had taken with Ollie, when they traveled to the Black colleges. A father and son together, discussing their current adventure and dreaming of possibilities for the future. If it hadn't been for those recruitment trips, perhaps Ben never would've reached the heights he did. Which meant that my dad and I never would've sojourned to DC to reopen his estate to keep his legacy—our family legacy—alive. One of the most important results of the Google commercial saga was that it made me believe, even more than before, in the story of my family.

Around the time the Davis Barracks building was dedicated, I began purchasing Ben memorabilia on eBay. When West Point had asked us what we'd like to place in the time capsule, I realized that we had very few personal items. Almost all of Ben's meaningful possessions are in the Smithsonian archives. So I took it upon myself to find whatever items I could and start a new family collection.

The more I searched certain time periods, the algorithm recommended other products to me, including vintage postcards; postcards were extremely popular in America back in the day. I felt compelled to purchase them, and I now have over a thousand Jim Crow–era postcards. They are so racist it was hard for me to believe they were real. As an example, one of them has a picture of a Black man chasing a chicken and saying, "Did someone say foul ball?" as a play on words with "foul" and "fowl." These postcards visually represent the kinds of stereotypes Ollie, Ben, and the Tuskegee Airmen had to contend with and overcome, just to have the same opportunities and respect as their white peers. Blacks have been disproportionately pummeled with brand equity challenges since day one.

Other items recommended to me for purchase on eBay were for Aunt Jemima. I remembered Aunt Jemima from my childhood because my mom cooked lots of pancakes and waffles. (You might recall that my dad is really into breakfast.) She'd say to me, "You know, she used to wear a bonnet." Then my grandmother Meredith would chime in with some comment. I don't have a lot of positive memories about the Aunt Jemima character, but the pancakes were delicious.

That one recollection sparked something in me, and I started to collect historical Aunt Jemima merchandise, including mammy dolls, newspaper ads, and magazine ads. As I accumulated dozens of items, I looked into the story and discovered a book with the provocative title *Slave in a Box: The Strange Career of Aunt Jemima*, published in 1998. I bought a copy to read and learn the brand's history. Aunt Jemima was created about twenty years after the Civil War ended and heavily relied on the mammy stereotype of an overweight, maternal Black woman who speaks in slang.

I became obsessed with this story, especially because I was working in advertising, the industry that had created this idea in the first place. I brought old Aunt Jemima ads that I'd purchased to the office to discuss them with coworkers. Obviously, I couldn't lug my entire collection of Aunt Jemima items everywhere I went, but I did carry around *Slave in a Box*, as a quick and easy way to capture people's attention and convey the message.

As a diversity officer at an ad agency at the time, part of my job was to ensure that our clients mitigated risk for anything that might be considered culturally insensitive. I was already working with PepsiCo on supplier diversity, through a platform I'd built out called One Sandbox. PepsiCo also happened to own Quaker Foods, and under that umbrella owned Aunt Jemima. This connection, albeit layered, could be an opportunity to speak to someone who was in a position to escalate the matter.

One of my contacts on the client side was a PepsiCo procurement leader. He invited me to a meeting at the company with a key executive—let's call him Dan Daniels. The morning of the meeting I showed up carrying a huge duffel bag. I hadn't mentioned to my colleague I was bringing this bag or about the conversation I had planned. If I had, he would've advised against it or even uninvited me to the meeting.

During the conversation, Dan Daniels mentioned an upcoming project that was a few years out, which would touch on diversity. It would be a movie called *The Color of Cola*, and it would explore the company's heritage. I told him that if the brand wanted to create movies about topics like diversity, they needed to do all they could to guarantee that nothing else associated with the company would undermine that effort.

"I think it's time to retire Aunt Jemima," I said.

Dan Daniels was a bit caught off guard. He assured me that all litigation related to Aunt Jemima had been resolved.

"Well, let me show you something," I said and removed a copy of *Slave in a Box* from my bag.

"Have you ever seen this book?" I asked him. "It's the biography of Aunt Jemima."

"What?" Dan said.

I then pulled all the items out of my duffel bag and spread them on the table. I had no idea how he'd respond, but I figured I only had this one shot.

I told Dan, "This is a ticking time bomb. Look at these dolls—they're all blackface. They're all mammies. All the good stuff going on around your brand is eventually going to go wrong because one day, this will become a problem, and we won't be able to control it."

Dan said that the way I shared the story put a different perspective on it, and he put in a call to have a meeting with someone about the size of the market share. He thanked me, and I left the room with his colleague who'd arranged the meeting.

Not surprisingly, the colleague was perturbed. What was I doing, walking into Dan Daniels's office and putting a bunch of blackface dolls on the table? I told him I was trying to shine a light on the subject and was giving them a chance for risk management.

Sometime later, I got a call from Dan, who consulted with me about putting together materials for a presentation to persuade the higher-ups to retire Aunt Jemima. I connected with some of his other colleagues, and a plan was created for a big meeting in February 2020, where Dan Daniels would pitch the idea internally.

When Dan called me after the meeting, I was hanging on his every word. He had been told that Aunt Jemima, as the face of a global brand, would never be retired.

Yeah—the face of a global brand who was never compensated for having her likeness plastered all over the place, yet made a fortune for the corporation who sold the products.

I heard the passion in Dan's voice and how devastated he was. As the call was about to wrap, he said there was one thing in particular that surprised him about the decision.

"What's that?" I asked.

He paused before saying, "The guy who rejected the plan was Black."

While race shouldn't matter, perhaps I was looking at this wrong? Did some people perceive her as a symbol of empowerment? That was a bridge too far for me, although that notion did broaden my perspective.

In spite of the rejection, we agreed that we needed to find a way to make it happen and turn the loss into a win.

That summer, following the murder of George Floyd, I received another call from Dan, who said our project was now on the front burner. By then, a new CEO had joined PepsiCo and an industry-wide debate had erupted around food mascots. Land O'Lakes had removed the Native American woman from their packaging, and Uncle Ben's was poised for a similar change.

Shortly thereafter, Dan phoned with the news that Aunt Jemima would be retired, and it would be announced that day. He thanked me and said that it never would've happened without me. I was humbled, but I had a feeling this was only the beginning.

I said, "Dan, I need you to do one more thing." I asked him to reach out to the other food manufacturers with mascots, like Eskimo

Pie and Cream of Wheat, to see if they'd consider retiring those. Today, I'm pleased to say that these and others have been rebranded.

Unbelievably, this situation ended up becoming a *Saturday Night Live* skit, with Aunt Jemima and Uncle Ben sitting in an office conference room being fired by corporate executives. And I was thrilled to call my mom and share the news.

Helping persuade PepsiCo to retire Aunt Jemima is one of my greatest professional accomplishments. There's not a lot of publicity around that moment due to its sensitive nature. Although I did do an interview for Yahoo! News that the Associated Press picked up. It was important to me to have one official "receipt" so that my efforts wouldn't be completely invisible.

Also, I wanted to tell the story, as a way to educate and inspire other people who wanted a road map for enacting these kinds of changes. I succeeded because of a combination of timing, luck, and industry savvy, and I'd learned the strategy of using the system to defuse the system from the very best—Benjamin O. Davis Jr. You have to be in the room to be the spark, and in this situation, I was proud to be that spark.

A major similarity across generations in my family is that we've all been better suited to speak to the establishment, knowing how to work with the system to forge lanes for ourselves and others. We've all been able to build bridges between two worlds, that of the privileged and that of the historically excluded, to convince other people that the greatest successes manifest when these two groups cooperate. Ben and the Tuskegee Airmen epitomized this dynamic: If the military brass would allow Blacks equal opportunities, that would result in a stronger military and improve the chance for victory. If Blacks agreed to participate in the Tuskegee pilot training program, they had a chance to prove they could be skilled pilots. Both groups benefited when they were committed to a common purpose of creating the most amazing military in the world.

Part of my purpose is to show people that working together *is* the opportunity. I consider myself a chief inspiration officer, because inspiration is my leadership modus operandi.

Granted, this journey hasn't been perfect or easy. Since moving to Switzerland, I've had to adjust to an unfamiliar country, culture, and industry. I had an opportunity to stay in the States and take the safe path. I could have bought a house in Connecticut, and tapped out of the career-climbing casino, so to speak. But I wanted to challenge myself with new experiences and gain a global perspective on a subject I've learned about solely in the context of American culture.

During my time in Europe, I've been able to travel to the places Ben lived and was stationed, from Europe to Africa to Asia. When retracing his footsteps, I unearthed the extraordinary things he did to find and live out his purpose. This inspired me to believe that there was no limit to the effort I could invest in my own purpose.

Ben was in my life for over twenty years, and I loved him. Knowing what he endured to achieve his dream continues to fuel me.

TWELVE

American

You can become your ancestors' wildest dream.

In the fall of 2022, I decided to visit the locations that house my family's archives, to see the items and understand how these museums came to possess them. I followed the process to schedule appointments and sent them a list of the specific boxes I wanted to access, so they could have them pulled and ready for me. I also hoped to speak to the collections' respective curators.

One of the challenges I've discovered in the years I've been advocating for my family, especially Ben, is that because we've been in public service for many generations, we don't own anything. Louis Davis worked for the government; Ollie and Ben worked for the military; my dad was a judge, and my mom was a teacher. So it's not as if they were entrepreneurs who had proprietary rights to any of their contributions. They all had access—which was important—but they didn't have ownership.

My first stop was Ollie's archives, housed at the US Army Heritage and Education Center at the Army War College in Carlisle,

Pennsylvania. As I drove there with my friend and RedCarpets.com business partner, William, I had a weird, unfamiliar feeling, one that was part combative, part excited, and part trepidatious. It seemed odd to me that a museum could own millions of dollars of my own family history, yet we don't. I didn't want to cause problems or conflict, but I was ready to have hard conversations about why they had everything. Why did they have it? How did they get it?

That day I learned that Elnora, Ben's sister and Ollie's younger daughter, had donated everything to the museum. They showed me the files of paperwork that included all the back-and-forth correspondence printed on pink paper in 1988. One option the family didn't take was to allow the center to digitize items and then return everything to the family. Though they're in the process of digitizing the materials so they'll be available online (a lot already is), they retain ownership of all the originals. And some of the original items can't even be viewed. According to the documents detailing the items in the collection, several boxes "are restricted for conservation purposes," meaning that they're unavailable to the general public or even family.[1]

Seeing personal items like baby pictures, high school diplomas, and letters made me upset and conflicted. It seemed wrong to me that such intimate family treasures were stashed in a museum in the middle of Pennsylvania, in a box and rarely seen or touched by anyone. But I conceded that we didn't know what would happen to the stuff if it wasn't in a museum. Would the family fight over it? Would it be lost or destroyed in a flood or a fire? Would it have ever been scanned and made available online, for anyone in the world to explore? I saw how organized and meticulously kept everything was and couldn't deny that the National Museum of the United States Army is best in class. They had obviously put a lot of time and effort into ensuring everything was safe and well-maintained.

One piece of information I wasn't able to find out on this trip is an estimate of the collection's value. We're talking about fifty-three boxes of everything from photographs to letters, postcards, and private journals. They even have some of Ollie's uniforms, from the late 1800s through the early 1900s, in a climate-controlled room. I'm not an antiques dealer or appraiser, but this collection must be worth millions. And it's a shame the family can't share in the value of it.

From Carlisle I traveled directly to Washington, DC, to view Ben's archives, which are housed at the National Air and Space Museum of the Smithsonian Institution. Much like Ollie's collection, Ben's archives are vast—168 boxes total—and cover his entire lifetime. Once again, I became emotional when I saw and touched the very personal items. The thing that stood out most to me was the many pictures of my dad as a child—pictures he'd never even seen.

On this visit, I was fortunate enough to meet the primary curator. She came out to speak with me and was as kind as could be. She told me that her husband was the person who went to Ben and Agatha's Arlington, Virginia, home to pick up the items in the archives. And that Agatha herself had greeted him and held open the door while the boxes were hauled out. She said that Ben's collection is the most complete of anyone from World War II at the Smithsonian—and the most thoroughly documented military experience of any Black soldier. They loved the collection so much they'd even written blog posts sharing recipes they'd found and prepared.[2]

When I heard about the extent of and their admiration for the collection, my next question was about the lack of visibility. Although I'd seen many of the items online, I'd been frustrated that none of the collection was on display anywhere. If it was so fantastic and complete, why was it hidden away?

I asked the curator if she could share with me the paperwork for the collection. She left for a little while to make some inquiries and receive a few approvals, then returned with the paperwork. I waited with both excitement and dread to see what she held in her hands, including a document I'll call "the note." The curator explained that they wouldn't normally share this paperwork, but they were making an exception because I was the executor of Ben's estate. She then placed the note on the table and talked me through its contents.

All of the items were indeed signed over by Agatha Davis, but beyond that, another message carried two instructions for the museum. The first condition was that the complete collection had to stay intact, meaning that it couldn't be broken up, and no portion of the collection could ever be loaned to other museums.

This was all news to me and quite jarring. Up until that point, I'd refused to believe that the way the Smithsonian acquired the collection was righteous or aboveboard. But in the calm, sincere voice of the curator, I heard from the documentation right before my eyes that my suspicion was wrong.

I hadn't even processed the first condition when the curator told me about the second. Agatha didn't include any negative articles or communications in the collection. Anything that could tarnish the family in the eyes of history or be used against them was removed. The curator shared that there was no documentation for a two-year period: the first two years they were at Tuskegee. My mind raced. Those years corresponded to the time Agatha had told Ben she couldn't have children. Was there some connection between their time at Tuskegee and the hysterectomy paperwork my mom and dad had found shoved under the mattress in the Arlington apartment?

Ben and Agatha must have endured so much pain and sadness during those years. They'd faced a lot of negativity in their lives, yet they didn't want that to be part of their legacy—they wanted to lead with joy.

The second part of the note, which was handwritten, revealed the most powerful thing I learned about this story. And I wasn't ready. If I hadn't seen it with my own eyes, I may not have believed it.

As the boxes full of donated items were being carried out, Agatha stood at the door and methodically looked through the contents. If she saw the words "African American" on anything, like a plaque honoring Ben, she took a sharp object and scratched out "African" every single time.

"We are American," she told everyone who was there from the museum. Throughout their lives, they'd been called Negro, colored, African American, Black—but they only ever wanted to be American, to receive the same equity as their white counterparts.

The curator went on to say that the Smithsonian had once tried to display some of the items, but because "African" in the phrase "African American" was scratched out, people thought the collection had been vandalized. So the museum stopped displaying it.

I was speechless.

The Smithsonian honored their request to abstain from using the term "African American" or from mentioning Ben's race unless it was warranted. The preference is even mentioned in the document that provides an overview of the collection: "In keeping with his sentiments, his ethnicity will only be mentioned when it has direct bearing on his career," and "At the request of General and Mrs. Davis the term 'black' or 'black American' is used in preference to 'African-American.'"

Agatha couldn't have known the consequences of her actions. Even if she had, I don't think she would've done anything differently. In that last chapter of their lives, it was a final effort to control the narrative. In a 1969 interview, Ben had once declared, "My attitude is that we should fight to be full-scale Americans. . . . To be an

American, to be accepted as an American."[3] He was adamant and consistent with this message throughout his entire life.

Society wanted Ben and Agatha to be thought of and remembered as African Americans, or even Black Americans. But in the eyes of history, they wanted to be remembered simply as Americans.

It has been an adventure through time and history that has made me more aware and appreciative of my love for my family and for America. But as thrilling as this has been, I recognize that the torch has now been passed to me. Because of the challenges that come with age, I can see the toll time has taken on my dad's life. He and I had stood side by side at every event to honor Ben, sharing the stage and our unique perspectives on the same story. Just like Ben and his father, my dad and I love spending time together, working toward the common goal of sharing about Ben's life and legacy.

Time and health seem to have compelled both of my parents to share even more information with me over the past year. The last time I was home they told me how I got the name Douglas versus their original choice, Davis.

In naming my siblings, they adhered to the style the previous generations preferred: Louis, Ollie, and Lawrence, my paternal grandfather, all named their sons after them. And Elnora, Ollie's youngest daughter, was named after her mother. My older brother's name is Scott, and he's named after my dad; my sister, Sonja, is named after my mother. However, eleven years after my sister was born, my mom was surprised to become pregnant with me. My parents debated what to name me, and my dad's first inclination was Davis, in honor of Ben. But at the last minute they decided to call me Douglas, which is my mother's maiden name. They both agreed that since I'm the one person in the family who has become so passionate about promoting Ben's story, it would've been fitting for me to be named after him.

The last time I was at my parents' home, my dad sat me down and told me, "You will soon be the patriarch. Soon your mother and I will be gone—and you must keep what's left of the family together." We hugged and cried, and I again faced the reality that whenever I see him, I don't know if it will be the last time.

A couple of years ago, my dad handed me an envelope labeled "Time." When I arrived back in Switzerland, I opened the letter he'd written to me. It began, "To my dear Douglas," and those words alone conveyed a level of sentimentality he rarely exhibited. I read on:

> *As one reaches the autumn of one's years, that's when we all come to appreciate the importance of time. Time is precious at any age, but becomes more important as time grows short. Most people don't come to appreciate the importance of time until just before it runs out. So they, including me, waste precious time in their early years doing silly things, unimportant things that are unlikely to be helpful to others in the future.*

I was grateful to receive the gift of this letter, which contained the wisdom of a father at the end of his life, encouraging me, his youngest son, to respect and honor the limited time I have on earth.

I do hope my dad's healthy enough to accompany me to the next milestone and ribbon-cutting for Ben, in June 2024. Recently, the US Air Force Academy Association of Graduates called to let me know they're building a new entrance and physical memorial for Davis Airfield. The installment will include full-scale replicas of Ben's red-tailed P-51 and of a Red Hawk, the modern red-tailed training plane, in honor of Ben and the Tuskegee Airmen. It's going to be a huge, celebratory occasion.

For an encore, I'm planning to write a letter to the president and his administration, seeking to have Ben receive a posthumous Presidential Medal of Freedom, the highest civilian honor in America bestowed by the president. Five years after Ben died, the Tuskegee Airmen received the Congressional Gold Medal, the highest civilian award given by Congress. Since Ben was unable to share in that, I envision my dad standing on the stage at ninety years old, while the Presidential Medal is carefully draped around his neck, as he accepts it on Ben's behalf.

All these years after my journey began, I can see how the quest never ends. In 2022, I received a Google Alert about something called the Davis Line. It runs down the middle of the Taiwan Strait, the body of water that separates Taiwan from mainland China. It's named after Ben because he identified the midway point by flying his plane down the straight. This occurred in 1955, when he was stationed as an Air Force commander in Taipei and tasked with establishing a stronger USAF presence on the island, to help stabilize the region after the Korean War.[4] As the highest-ranking military official on the island, Ben worked closely with Taiwanese leaders, including Chiang Kai-shek, to develop a defense strategy against any future attacks. The Davis Line "was designed to keep military aircraft from both sides of the strait at a safe distance in order to prevent miscalculations from erupting into conflict."[5] It had popped up in the news because of rising tensions between China and Taiwan. For decades, the Chinese military had only violated the unofficial border a few times, but in the past couple of years, they've entered Taiwan's space hundreds of times.[6]

These kinds of revelations blow my mind. Every day is a new opportunity to discover something about my family, about American history, and about myself.

Discovering where I came from has filled me with endless optimism. Just because the story has been invisible doesn't negate that it exists—that it is powerful and is still impacting my life, and even the world. Black history is American history—even that which has been invisible until now.

EPILOGUE

Inspired to Greatness

The morning of Saturday, December 17, 2022, was a blustery day in Kitty Hawk, North Carolina. Bracing myself against the intense gusts of wind made me understand why Orville and Wilbur Wright chose that location for their historic first flight. It was unbelievably cold, and it evoked an image in my mind of that frigid day in January 1873, when Louis Davis rode in the carriage during President Ulysses S. Grant's second inauguration parade. In some ways, that day was the beginning of my family's story. And this day felt similarly momentous to me.

The First Flight Society, a group founded to preserve the legacy of the Wright brothers and promote aviation and aviation education, had invited my dad and me to speak at their annual event. In addition to commemorating the one hundred nineteenth anniversary of powered flight, they were celebrating the seventy-fifth anniversary of the United States Air Force. My reason for being there was the third source of excitement: they were honoring Ben by inducting him into the Paul E. Garber First Flight Shrine, with a large commissioned portrait that would be unveiled on the most important date in flight. The painting will be featured at the Wright Brothers National Memorial Visitor Center for a year.

In discussing the portrait, the organization told me that they wanted to label it something like "Air Force's First African American General." Based on what I'd learned about Ben by that point, I knew that wasn't going to fly. At all. I suggested it be labeled "Father of Black Aviation." When they asked why, I said because he was basically the Black Howard Hughes—an amazing innovator with unsurpassed aviation knowledge, although his genius wasn't fully appreciated in his time. They may have been reticent about this label because that title is typically used for Chief Anderson, who'd famously flown First Lady Eleanor Roosevelt as his passenger in Tuskegee. In the end, they went with "Black Aviation Pioneer, Tuskegee Airman, American."

The Wright brothers took their first flight on December 17, 1903. The next day, December 18, the front page of the *Virginian-Pilot* declared: "Flying Machine Soars Three Miles in Teeth of High Wind over Sand Hills and Waves on Carolina Coast."[1] Nine years later, on December 18, 1912—the same date the news broke about the Wright brothers' successful flight—Benjamin Oliver Davis Jr. was born. I love the feeling of knowing that the energy of the universe brought Ben to us on that special day. It's as if his path was always meant to be.

Two of the speeches given that day stick out in my mind. The first was by Maria Tinker, the supervisory park ranger of the Tuskegee Airmen National Historic Site. Maria graduated from Tuskegee Institute and is exceptionally knowledgeable about the Airmen's history. She spoke of how much Ben inspires her and quoted from his autobiography. It was wonderful to hear from a young person who not only knows about the family but also strives to preserve Ben's legacy, along with that of the Tuskegee Airmen.

The person who presented the keynote and introduced me was four-star Air Force General Mark D. Kelly, who leads the Air Combat Command, a vast global operation with more than 1,000 aircraft, 200 locations, and 100,000 military personnel worldwide. He provided a concise and powerful overview of Ben's military accomplishments and his profound impact on the Air Force, the military, and the nation.

The closing speaker slot was reserved for the family. Usually I'd write up some notes in advance, but since two others were speaking about Ben before I was, I'd decided to just speak from the heart. As I took the stage, my eyes scanned the crowd of bundled-up attendees. It was a complete surprise when I saw the distinguished red jackets of three Tuskegee Airmen. Pride and joy warmed me from the inside out. While I knew they'd been invited, seeing them in person, voluntarily bearing the cold, heightened my appreciation for them having shown up yet again to honor the man who had led their way. These men are incredible.

One of the Airmen, Sergeant Thomas W. Newton, who's in his nineties, had served under Ben's command for three years. He later told me that seeing Ben inducted into the Wright brothers hall of fame was something he needed to witness, and he wouldn't have missed it for the world. I was so touched when he saluted Ben during my speech.

Any time I speak on behalf of the family, I lead with some version of what I said that day: "If [Ben] was still alive today—he really didn't like speeches. I can assure you he would just come up here, he would thank the Airmen, he would thank the Air Force, which was his family, and then he would walk right off. He was a man of few words."[2] But I'm not Ben—I love to talk, and I love getting deep into my family story.

After the ceremony, I went over to the First Flight Society's luncheon, where I was going to speak again. This day marked the first time I had two opportunities to share the story of the Invisible Generals with audiences that reflected America itself—full of the diversity that makes this country so special. This also was the first time I publicly shared the news of this book, and that it would be released the following year.

The story touched so many people that day. It makes me smile with gratitude when I remember how great it felt to be there, connecting Ben's legacy with America's legacy. I heard a lot of positive feedback about my storytelling from people I chatted with at the events, and I received uplifting and encouraging emails from people afterward.

That night, after I flew into New York City and drove to my parents' home in Connecticut, I felt satisfied in the quest I had undertaken to discover my family's legacy, and secure in the knowledge that this story is just as powerful today as it ever has been. It has immense potential to inspire not only those of us living today but also those in generations yet to come.

PART II

BECOMING A VISIBLE GENERAL

Strategies to Seize Your Destiny

I love to tell the story of the Invisible Generals. My friends jokingly tell me that I'd attend the opening of an envelope if it meant an opportunity to tell the story one more time.

But I think the story is powerful because it stirs conversations about history, it unites and inspires people, and it is thought-provoking. It's an intimate family story that also contains universal themes and struggles, so it resonates with people from all walks of life. And in many ways, the story has strong parallels to the times we're living in.

After I tell the story, I often ask the people listening if they've ever talked to anyone in their family about their story. Where did your mom grow up and go to school? What was your dad's dream occupation when he was a child? What was your grandfather's biggest struggle? What was your grandmother's proudest achievement?

Even though I myself had never bothered to ask these types of questions about my own family until that fateful *Red Tails* screening, I'm still amazed—and reminded when I inquire—that most people haven't turned to the generations before them and said, "Tell me your story."

I know what it's like to be in the dark about a legacy you never knew you had. That's why it hurts even more when I learn of stories that are at risk of being lost forever, as our elders age, die, and take those histories with them.

Be confident in the power of your stories. It took some time, but I learned to be confident in mine. We are, after all, the progeny of survivors.

If I can leave you with one simple idea from the story of the Invisible Generals, it's this: begin your own journey of discovery in your own family tree, and in your own community.

No matter what your roots are, you have the power to become your ancestors' dream descendant. These steps are your playbook for that adventure of a lifetime.

Discover Your Generational Collateral

We're the receipts of our ancestors' journeys. For some of us, this means our lives have been infused with support, love, and stability. For others of us, our lives severely lacked these positive elements. Regardless, we can foster a sense of empathy and grace for previous generations, recognizing that even if our family's past isn't ideal, we have the power to change our present and thus the future. Our success isn't determined by what we were or were not given—it's determined by how we use what we have while we're here. Each resource you have may be an essential tool in strengthening your family and community, and our world.

Be Curious and Ask Questions

Family stories can become valuable generational collateral of their own, but you have to get people talking. Some of my favorite and most vivid memories are of the times I've spent with my dad over the past ten-plus years—even the time he laughed in my face when I vented to him after the *Red Tails* screening. Based on my own experiences of struggling to extract the story from my family, I understand that sometimes people don't want to talk about the past. If there's a lot of hardship or trauma in your family history, that's perfectly understandable.

However, I encourage you to be persistent. As I discovered, sometimes you just need to crack open the door to conversation and, over time, people typically become more willing to answer questions and tell you more about their life and about other relatives. Each new piece of information you learn could be a key to the next step.

Think about the story you're telling about previous generations in your lineage, and therefore about yourself, through your words or actions. Is it a story of love and hope? A redemption story about overcoming challenges? Be honest about the struggles. Future generations can learn just as much from past mistakes as they can from past triumphs.

Part of preserving your family story and honoring the legacy you've inherited is an obligation to tell the story and speak power into it. In reclaiming our stories, we can shape and mold not only the present but also the future. Don't let the world tell you who your family is or who you are—you control the narrative.

Recognize and Say Thanks

Most of us have people who helped us in some way as we worked toward our life goals. Of course, parents or guardians are usually the obvious providers of our basic needs—food, clothing, shelter, and love. But others may have contributed too. Maybe someone provided a financial lift, like an aunt or uncle who saved money to help you buy your first car. Or perhaps it was a connection or opportunity, such as a community elder who gave you a job. Maybe it was an investment of time and energy, like a grandmother who provided childcare while you took college courses at night to finish your degree.

Yet, how conscious are we of these investments others have made in our lives?

For my part, I appreciated everything I'd been given in life, especially by my parents. However, until I started learning more about my family story, I didn't understand the depth and breadth of their sacrifices, and everything previous generations had endured, to grant me the privilege of living the amazing life I have today.

So one of the first and most important steps of building your own generational collateral is *recognition*. In addition to your relatives, have you ever considered the support structure you had when you were growing up, such as the teachers who taught you, the cafeteria workers who fed you, the bus drivers who transported you, the coaches who motivated you?

Take the time to think of all the people who have helped you down the path, even when they weren't obligated to do so. Make a list of names. Then, start thanking these people, letting them know you appreciate what they did for you. And don't just send them a text or post a comment on their social media page. Pick up the phone and call them or send them a handwritten note. Better yet, offer to take them out for coffee or lunch and tell them in person. Also, be specific about what they did for you and the positive impact it had on your life.

Quite often, people think their acts of kindness and generosity go unseen. We have the ability to change that, to start or continue the process of crafting a story of gratitude for them.

Pay It Forward

Once you've taken the time to evaluate and appreciate all the ways other people have given you a hand up in life, you can start brainstorming ways to do the same, whether for people in your family, in your community, or both. What kind of generational collateral can you provide for others?

Though financial contributions might be the first thing that comes to mind—for example, starting a college fund for a young cousin or donating to an after-school program for underprivileged youth—don't limit your ideas to acts with dollar signs attached. Besides, the reality is that not everyone can provide money, stocks, bonds, and other tangible assets.

Truthfully, some of the most valuable assets you can provide are intangible, like knowledge, experience, and encouragement. If you can't start a college fund for that young cousin, you could edit their college entrance essay or help them study for their SAT. If you can't donate to the after-school program, volunteer to be a tutor or mentor. Every contribution counts. We serve ourselves and future generations best when we share whatever resources we can; when we educate, inspire, and nurture; and when we stay positive and hopeful.

Honor Your Family Legacy

Intentionality is a major factor that determines how effective we are in honoring the legacy we've inherited. Our society is structured in such a way that deep thinking about our decisions and allowing ourselves to consider and weigh options sometimes seems impossible. However, impulsive and short-term thinking and actions undermine our ability to be innovative in our personal and professional lives.

If your days are crammed full of bingeing shows on Netflix and scrolling through your Instagram feed, are you going to have time to write that book? To daydream about what you want your life to look like in the next year or five years? To work on the business plan for the company you want to launch? To invest the time and energy in discovering, preserving, and promoting your family story?

Any time I endure injustice or think about giving up, I turn back to my family and think, *What would Ben do?* One thing I do know for sure: he'd focus on the long term and keep his eye fixed on the ultimate goal, whether that was attending West Point so he could become a pilot or becoming a four-star general.

Dream of a Million-Dollar-Star Scenario

Try to identify at least one ultimate goal, which will be your own million-dollar star. Then promise to meet that goal. You don't have to share this promise with anyone else—you can make a promise solely to yourself. If your family is involved in what you're doing, you can collectively promise to reach for that million-dollar star to secure your family's legacy.

Personally, I think my dad's million-dollar star was becoming a judge. For me, it would be an appointment to sit on the board of directors of a publicly traded company. All of us have personal goals, such as saving a certain dollar amount for retirement, paying off our mortgage, or graduating our children from college. If you're an entrepreneur, perhaps your million-dollar star is hiring your first employee or leasing your first office space. In corporate America, you land in an executive suite in a swank corner office. In academia, it can be earning your PhD or getting tenured.

Within the context of your family legacy, here are a couple of examples: Let's say you want to have an elementary school renamed for your grandmother, who taught reading at the school for decades. Your million-dollar star would be the moment you and your family pull the cloth off the newly installed school sign, revealing her name. Or perhaps you were adopted and want to start an organization that supports parents of adopted children. Your million-dollar star might be the day the organization graduates its one thousandth student from your pre-adoption parenting program. The possibilities for million-dollar stars are limitless, because they can represent almost anything and be adapted to anyone.

Ollie in uniform as second lieutenant of the 10th Cavalry, US Army, 1902. (From family collection)

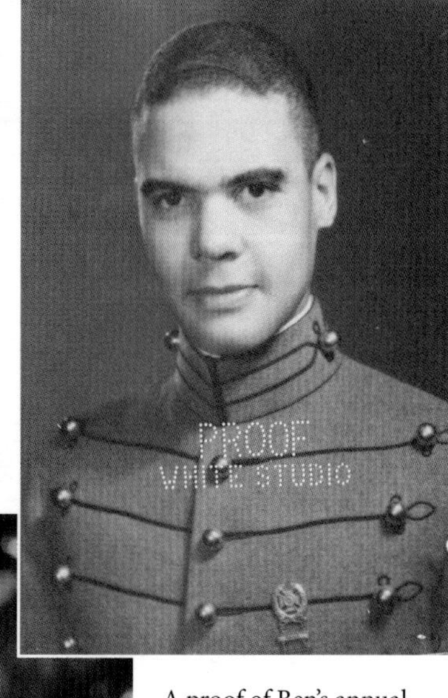

A proof of Ben's annual class photo at the United States Military Academy at West Point, 1936. (From family collection)

Ollie and Ben on West Point graduation day, 1936. Upon Ben receiving his commission as a second lieutenant, the father and son were the only two Black US military officers until the start of World War II. (From family collection)

Ben was frequently photographed with his wife of over sixty years, Mrs. Agatha Davis. This picture was taken in 1943, at Selfridge Field, Michigan. (From family collection)

Ben enjoying a smoke while posing with his P-51C Mustang "By Request" in Italy, circa 1944–45. (© Chanute Air Museum, All Rights Reserved. Used with permission.)

Young Larry Melville sitting in the cockpit of one of Ben's planes, circa 1948–49. (From family collection)

Larry posing with model airplanes, August 1958. His poor eyesight prevented him from becoming a pilot, so he built model airplanes to satisfy his passion. His plane is #6, third from the right. (From family collection)

On the steps of the White House in 1948, President Harry S. Truman presents Ollie with a scroll commemorating the retirement of the military's first and only Black general after fifty years of service. Ollie's second wife, Sadie, stands by his side. Ben and Agatha also attended, although they're not pictured here. (From family collection)

President Richard Nixon meets with Ben, whom he'd appointed as director of Civil Aviation Security for the Department of Transportation, 1970. During his tenure, Ben would help increase commercial airline safety. (From family collection)

Ben walking through a metal detector device at the Conference on International Air Transportation Security, 1971. He led efforts to create what is now known as the Transportation Security Administration (TSA). (From family collection)

President Gerald Ford meets with Ben in 1974. (From family collection)

Ben stands in the Oval Office with President Jimmy Carter in 1977 to unveil the iconic 55 mph speed limit sign. Known as "Mr. 55," Ben helped identify a speed limit that would both save lives and conserve gas. (Jimmy Carter Presidential Library)

Doug visiting with Ben in Washington, DC, while Doug was on break from college at Syracuse University (1994–97). (From family collection)

Ben, Larry, and Doug pictured moments before heading to the White House grounds for President Bill Clinton to pin on Ben's fourth star in 1998. This was the first and only time Doug saw Ben in uniform. (From family collection)

A family gift from President Clinton, a signed photo of the forty-second president and Ben's sister Elnora pinning the fourth star on Ben's uniform in Washington, DC, on December 9, 1998. (From family collection)

At Arlington National Cemetery, Doug places a rose on Ben's casket, as a final goodbye gesture. Ben passed away on America's birthday, July 4, 2002, and was buried alongside his beloved wife, Agatha, who'd died only a few months prior. (From family collection)

I encourage you to do everything in your power to keep the promise. Even if you don't attain that million-dollar star in your lifetime, you will have laid the groundwork for future generations to pick up where you left off.

Find Your Community and Create Your Own Monuments

The story of the Invisible Generals involves actual, constructed monuments. You probably aren't navigating the construction or destruction of external monuments. More often than not, the "monuments" we contend with in our daily lives are those we build together. On the positive side, they're the confidence, self-worth, and good health we enjoy that allow us to thrive. On the negative side, they're the inner critic, limiting beliefs, and other mental, emotional, and sometimes physical obstacles holding us back.

This is why it's important to find like-minded people who align with your purpose and goals. When I became a diversity officer, I didn't have a "monument" to look at for inspiration. And I had no clue what the job entailed. That's when I found Tiffany R. Warren, who showed me the way and invited me to help her form ADCOLOR. This organization helps people in the creative industries not only rise up but also reach back and allow the tide of success to pull others in and along for the ride. Her guidance and the support of the community have been and continue to be important to me.

By including me in the movement she was building around diversity, I was able to have someone who could help me answer questions like, what kinds of tangible goals could or should I aspire to reach? And how would I meet those goals?

Rob Schwartz was my CEO when I worked on Madison Avenue, and he brought me into his group of creatives and taught me how to think like a creative CEO. He helped me understand the monument of Madison Avenue, which is a foundation of commerce in both the United States and globally. This was eye-opening to me, and under his guidance, I was inspired to learn about the craft of creativity and how to make a "monument vision board" displaying words and images that represent your hopes and dreams, and how to get there.

Articulate Your Purpose

I'm sure hundreds, if not thousands, of stories are out there waiting to be told. And sometimes the best way to honor that story is by learning it and sharing it with the world. I feel driven by my ancestors and their energy—they are my fuel. What drives you?

Everyone's purpose is different, and the world doesn't need you to live someone else's life: it needs you to be the first and only *you*, whether you're here to design the world's most eco-friendly skyscraper, invent a new plastic alternative, or launch a first-of-its-kind business or nonprofit.

Sometimes we find our purpose by accident. Maybe you're learning about your family story and discover a treasure trove of vintage family photographs. You see that most of the photos aren't in great condition, so you start researching ways to restore them. That could ignite a passion in you that leads to a career you never envisioned, such as a photo archivist or museum curator. You can then leverage this career and the resulting professional connections to further your family legacy.

I believe that when you've identified your purpose and live in it, you move differently. You're more committed, determined, focused, and consistent. Everything in your life falls into place and aligns with that purpose—how you spend your time, where you spend your money, where you work, how you raise your kids, and so forth. Living within your purpose is something that changes both your mindset and your behavior.

Enhance Your Brand Equity

Early in my career I was told, "If you want to excel at something, you have to prove you can do it on your own." In my professional life, I help companies shape and present a unique brand through their overall image, products, and objectives. In my personal life, I do this same work directly for my family, which I feel makes me a more capable diversity officer—which then allows me to continually invest in those I love. It's a reciprocal relationship, a natural connection between career and cause. In both situations, I serve those who have been historically excluded, whether from corporate environments or the history books, and I convey messages through marketing and storytelling, two skills I've developed over the years.

Leverage Technology to Establish or Sharpen Your Presence

Technology has played a key role in my discovery of different parts of my family story, in my learning more about history, and in my ability to connect to people who hold pieces of the story, no matter how small or large. I've used everything from Google, LinkedIn, eBay, and Amazon to pull everything together over the years. Additionally, I've applied this same technology to building both my personal brand and the Invisible Generals brand.

I suggest creating a website or blog that people can follow, to see your work and stay up-to-date with your activities. A website might be a good option if you already have a lot of information and visual assets you can include. However, a blog might be beneficial if you're just starting the discovery process, so you can document the journey in real time. This advice might seem obvious or rudimentary, but take a new headshot and write an updated bio to use for your website and socials. People need to be able to attach your face to your message and your family story too. Finally, set a Google Alert, so you know when you or your family are mentioned online. That way, if you find out something is happening, you can get involved. The information you glean from Google Alerts, such as news articles, can also be linked on your website, to keep its content fresh.

Differentiate Yourself

For several years, I've been a lecturer at Stanford Graduate School of Business, in Allison Kluger's "Reputation Management: Strategies for Successful Communicators" course. I consider *reputation* a synonym for *you*. Your reputation is everything about you—the way you look, the way you talk, your personality, your energy, your style, etc. It's the only thing that precedes you when you walk into a room.

One of the most critical characteristics of building a successful reputation is differentiation. To cut through all the noise that exists in today's world, you must establish yourself as a unique voice and force. A caveat here is that while you want to be unique, that doesn't equal complex. Strong brands have attributes that are easy to recognize and remember.

Differentiating yourself or the cause you're championing doesn't need to be difficult or complex. Start with the basics.

Some attributes will be imparted to you, and right or wrong, these are often based on your appearance and actions. The Tuskegee Airmen are known for their bright red blazers, which is a positive visual brand based on appearance. This is something I've subtly adopted by incorporating an element of red into my outfit when I speak at events on my family's behalf.

In your life, maybe you're the first one in the office every morning, so being an "early bird" becomes part of your professional persona. If you're always late to meetings and events, then part of your persona might be "the one who's never on time."

If you want to be remembered for something, repeat it three times in a room. Say you want people to associate you with timeliness. When you walk into a meeting, tell the other attendees that you wanted to make sure you were on time because being on time is important, and you can never be on time too often. Then, when you're mentioned in future conversations, even if they don't recall your name, they'll think, *Oh yeah—they're the person who's always on time.* This approach can also serve as an antidote to attributes of your brand you want to recalibrate.

Harness the Power of Perseverance

Throughout my lifetime, from both previous generations and my own experiences, I've been taught an infinite number of lessons about how to persevere. One of the most impressive examples was Ben's multidecade wait for his fourth star. I can't even imagine waiting a mere year or two for something I felt I'd earned or deserved. That kind of long-game mindset is something I've learned a lot about, from the years I've already invested to the many years ahead, as I continue to pursue opportunities to tell the story and secure my family's legacy.

Spending the past decade-plus piecing together my family story has taken a level of patience I didn't even know I had. I've had to learn where to locate useful sources during the act of discovery and how to network. I've sharpened my ability to phrase effective questions that elicit the information I need. And I've fine-tuned the way I communicate with people about the story.

When you begin this process, you might become frustrated or discouraged because people are unwilling to share information or answer questions, or you continually hit dead ends. I encourage you to stay the course, because you never know when one conversation, one seemingly random piece of info on the internet, might be a key that unlocks doors and opportunities.

Ollie and Ben often talked about how the Grand Canyon was once nothing more than a trickling river. For fifty years, Ollie was being a pioneer for his son, as well as being one for the Black community and the nation. Ben followed in his footsteps, spending decades blazing trails for others. You, too, can take actions every day to advance people, communities, and culture in incremental ways. And the results of those actions can carve out a new, different world.

Expect Setbacks

Without preparing yourself for setbacks, it can be easy to get derailed when things go wrong. But if you enter into a situation with a mindset that something can and will go wrong and strategize how you'll navigate that, you'll be better able to overcome and move past that obstacle.

For instance, when Ben was working to improve air travel security, he had to coordinate with lots of different stakeholders, from privately owned airports to various government agencies. He had to anticipate their objections and effectively address their concerns, operating like a military strategist in an office setting. If he'd gone into that process thinking, *Everyone's going to do everything I suggest because they want to keep passengers safe*, he would've been shocked and dismayed when people refused to cooperate. Their resistance might have made him give up. But his life experience, from West Point onward, had taught him that conflict and disagreement were inevitable, and the best way to handle it was to be prepared.

With your family story, know that you'll sometimes hit roadblocks, wind up at a dead end, face criticism, and get discouraged. And that's okay—it's all part of the process. In the low times, it can be easy to forget our little victories. Whatever actions you're taking, document the changes, both small and big, that have been enacted.

This has the added benefit of helping you notice patterns in your activities. You can see what's working or course correct if certain actions haven't proven as impactful as you'd hoped.

Seek Support

Investing time and energy in any cause can be exhausting—mentally, emotionally, and physically. Thousands of stories are out there about the blood, sweat, and tears that inventors and entrepreneurs have poured into their work, undoubtedly with more failures than successes. Without the support of family, friends, and others alongside you, you might be tempted to quit.

If you're peeling back layers of your family story, the things you discover might challenge your notions of what you've always believed to be true. Some people in your family might resist your efforts. Having the support of either friends or like-minded relatives can be critical to your ability to persevere. Being surrounded by people who speak truth, encouragement, and love into your life can make all the difference.

Ben heavily relied on his father, stepmother, and wife. Agatha in particular was his rock through everything. Likewise, I've been blessed with family members who support what I'm doing—my mom and dad; my brother, Scott D. Melville, and sister, Sonja-Lisa Melville; my cousins Trey Burns, Sarah Fornaris, and Danielle Burns Wilson—as well as innumerable friends and colleagues who have been on this ride with me, through all the ups and downs.

As another form of support and encouragement, create some kind of visual record, whether it's a website, a presentation, or a folder full of photographs from events or thank-you cards from people whose lives you've impacted. Having these kinds of visual reminders can motivate you and give you tent poles in your career that help you better tell your story and remind you where you've come from—especially when the journey becomes challenging. Celebrate your progress, knowing that each success is a brick you're laying in the foundation of your legacy, which future generations will be able to build on.

Stay Focused

In today's world, particularly in America, most of us have more opportunities than we could even begin to pursue. Where we go to college, where we live, where we work, how we spend our free time—even this handful of choices presents myriad options we need to weigh against our purpose.

When I look at my family, it's easy to see how any of them could've ended up on a different path. Ollie could've become a preacher or a government employee if he'd followed his parents' wishes for his life. Ben might never have become a pilot and general if he hadn't gotten into West Point. My dad could have entered a completely different field, if not for his fateful visit with Marcus Tucker or for Ben's influence on his life. My mom knew she wanted to be an educator and never deviated from that path.

I had a similar experience to Ollie when I decided to become a chief diversity officer. Though I never had any misgivings about my purpose, some people close to me did. My dad felt that trying to make companies equitable was a waste of my education and time, and that speaking out on the third rail of race and equity at work might hinder my career. Surprisingly, one of my biggest supporters was William Sidarweck. We'd built a business together, but I chose to transition away from it to pursue my purpose. His encouragement reinforced my decision.

When you live in your purpose, infinite opportunities may present themselves. Assess how each one aligns with your mission. It may be a worthy cause or something you've wanted to try or that excites you, but consider the opportunity cost involved, and keep both the long and short term in mind. Make sure that the undertaking won't take you so far off course that you'll have fewer resources to commit to your mission.

Serve and Inspire Others

A theme in my family's story is the way they cleared the land and paved the road for those who followed. It's like a team of railroad workers who labored in the blistering sun to lay track, so we could zoom along in our climate-controlled passenger train. It's a *service mentality* versus a *selfie mentality.*

Another important lesson I learned from the generations before me is that you can't allow the freedom to live an ordinary life make you complacent. There's always work to be done.

Let Go of Expectations

It's almost impossible to lead and inspire others when we're trying to impose our will on them, which I've wasted a lot of time doing over the years. Sometimes we try to convince people to conform to our vision of what they should want, be, or do. However, relinquishing control of other people tempers judgment, because you recognize they're also managing things in their life that they have no control over—things that may be overwhelming them. For instance, maybe they exist in a dysfunctional familial relationship they've not yet figured out how to navigate. Maybe they are constrained by a childcare situation, or an undisclosed medical need. It does not serve us to assume the mentality of those around us before we have been shown that mentality ourselves.

Another benefit of giving up control of others is that you'll be less inclined to blame them when things go wrong. Even though they might have contributed to the problem in some way, only *you* control what happens next. When Ben was silenced at West Point, he could have allowed bitterness and anger to take root in his heart and mind. However, he wisely chose to rise above it all, taking ownership of and mastering his mind and emotions.

The same happened for me, when the housing and stock market crashed in 2008, and my time with Magic Johnson had come to an end, with no job prospects in sight. I wasn't upset about not being asked to stay on with him, nor was I upset that no one came calling for me. I used that time and energy to fuel myself to start my own business and teach myself how to move past my own blind spots. That led me to become a better learner and businessperson.

These ideas apply in your profession too. You likely have little to no say in your supervisor, coworkers, duties, and so forth. You might not have any recourse when others' actions cause problems. But you do have control over both your *preparation* and *performance*—as Louis, Ollie, Ben, and my dad all demonstrated. Those two factors are crucial for leaders, and no one can ever take them away from you.

Lean into Your Leadership Style

Seeing the styles across generations was revelatory. Ollie and Ben were servant leaders by virtue of being military officers, yet their primary leadership style was performance-based. Performance was Ben's big thing, and his father had undoubtedly drilled that into his head. He'd tell me the same thing he told the Black college recruits: "Don't worry about all the scholarly people making all that noise—it's about getting the job done." He was motivated to excel at everything, convinced that sublime execution on any task would secure more opportunities for himself and those who'd come after him. That's what inspired him, and that's what inspires me.

If you're not sure what kind of leader you are, invest the time to find out. Dozens of free leadership-style quizzes exist online, which can help you identify your style. Of course, your leadership approach might shift a bit based on the environment (for example, you might exhibit a different style at work versus at home or at church), but in general, I think we tend to have a default way of operating.

After you learn more about your leadership style, it can be a pathway to inspiring other people. I was inspired by Magic to be more relational and not to focus solely on my own wants and needs. Ben's performance-based style lit a fire in me to be the best at anything I set out to do.

Recognize Service as a Universal Superpower

As soldiers, service was a way of life for Ollie and Ben. When they traveled across the country visiting Black colleges and universities, the outcome of their investment of time and energy was a big question mark. Nevertheless, they were committed to the mission and the students they encountered. Ollie and Ben trusted that, regardless of the end result, being able to uplift these young Black men was valuable in and of itself.

I think we often make service much harder than it needs to be. Sometimes even a seemingly minor act can have a major impact, and we shouldn't underestimate our ability to have a positive effect in any area of our life. Although we can never be 100 percent certain the seeds we've planted—whether in a person, project, or organization—will bear fruit, we can still learn a lot from the process itself. Be on the lookout for both small and large ways to serve others, from leaving an encouraging note on a coworker's desk to chairing your favorite organization's annual fundraising gala.

Serving others is a choice, but it's one we can make each day with intention. Whether you're serving in the context of securing and furthering your family legacy, similar to what I'm doing, or you're serving another person or a cause, it's almost impossible not to be inspired in the presence of someone who's chasing their passion and walking in their purpose. *Be that someone.* People will get excited about what you're doing and want to support you.

Raise Your Voice

Various experiences in my life, including becoming an advocate for the Invisible Generals, have taught me that in our own way, in our own communities, we can leverage our skills and use our voice to help carve out space for something bigger. I first had to find *my* own voice in order to create a pathway for those whose voices were rendered silent and thus were invisible. I also had to learn how to be an effectual advocate for causes and people. As with any new skill, it took some trial and error, but over time I developed a method for having direct yet nuanced conversations that compel people to action.

In my life, the professional and personal are integrated and reciprocal, so regardless of the environment, I'm able to bring 100 percent of myself to any situation or environment. Numerous experiences have reinforced this synchronicity.

Assess Your Skills

Sometimes American culture seems to encourage us to be generalists who know a little bit about everything and can do many things at an average level. However, I encourage you to become a subject-matter expert of your family story and honestly assess your strengths and skills. Find one or two things you excel at. You can then become a master at applying that to your advocacy work. The possibilities for ways you can use your skills and gifts are limitless, but here are a few examples:

SKILL/GIFT	ACTIVITY
Writing	Pen an op-ed for your local newspaper, or for your hometown paper, sharing what you've discovered about your family history.
Singing	Compose a song about your family story and perform it at a local open mic night.
Event planning	Coordinate a family reunion.
Web designing	Create a website that provides an interactive timeline of your family story.
Public speaking	Reach out to relevant podcasts and offer to be a guest on an episode or two.

Fine-Tune Your Storytelling

The storytelling aspect can't be undervalued. You must be strategic in making the narrative relatable. You're telling the same story repeatedly, and you need to position yourself as a credible expert as you speak with people.

When you're in discovery mode and beyond, be on the lookout for something you see or hear that might provide either a kind of "hook" for your story or a new direction you can take.

As you're piecing together your family story, identify and share the salient bits with others. For example, did you learn something that completely blew your mind, which you can't wait to tell someone about? When you begin telling your family story, are there parts where people's eyes light up, and you can tell they're fully engaged? If you're hanging out with a mixed crowd of people, does one of your friends say, "Oh my goodness—you *have* to tell them about X"? Answering all of these questions will help you home in on what resonates most with people and refine the narrative.

It's also important to understand and utilize the power of visuals in your storytelling. Think of how powerful a single image can be: the soldiers raising the flag at Iwo Jima or civil rights protesters being fire-hosed. When you see those kinds of images, even if you don't know the story behind the photo, the photo itself tells a story.

With the Invisible Generals, I'm blessed with tons of visuals, thanks to the seemingly bottomless archives related to Ollie and Ben in museums, archives, videos, and photos. But what if you don't have any photographs related to your family history or cause? That's an opportunity to use resources like historical or stock photos, and color palettes, text, and other elements to create a mood or mindset related to your project.

Prioritize the Effort

In our world of jam-packed schedules, it's easy to let seemingly more pressing or important matters push this advocacy work off your calendar. Remember that you're investing time in other people, which requires commitment. For instance, you can say, "I'm blocking out two hours for this every week." And don't just say it—put it on your calendar.

To maximize your time when you sit down to do the work, have a plan in place. Who do you need to call or email for more information? What websites do you need to access for research? Do you need to meet up with a friend to practice telling the story? Do you need to organize your notes? Try to identify at least one main goal you want to accomplish in each session and strive to complete it.

Use Optimism as an Operating System

The preceding generations in my family never stopped believing in the promise of America and the possibility of achieving the American dream for them and their children. They also took ownership of their lives—their decisions and actions, and the resulting outcomes. They understood the power in claiming their agency and refused to be victims. These lessons are some of the most poignant and meaningful ones I've learned.

- *Surround yourself with optimists.* This probably seems obvious, but some relationships in our lives tend to exist by default. That overbearing parent or incessantly complaining friend won't be the best traveling companion for the difficult legs of your journey. Since we don't necessarily get to choose all the people we're surrounded by (e.g., family and coworkers), it behooves us to build a community of positive, like-minded people who can support us.

- *Focus on the positive.* This might seem obvious too. However, I know from my life how easy it is to fixate on our own and others' mistakes and shortcomings or the things that aren't the way we want them to be. In either the present or future, we can mindfully examine a situation from every angle, seeking the good.

- *Guard your mental health.* On any given day, anyone who spends only a couple of minutes reading news headlines is likely to experience some anxiety. Those of us in the younger generations are privileged to have enhanced awareness of and access to mental health resources. Take advantage of anything and everything you need to maintain your internal well-being.

- *Avoid comparing.* Your path is your path, and no one else's. You might look at yourself and say, "I should be much further along than this." Recognize that someone ahead of you on the path is looking back and saying, "I wish I had gotten that far at that point in my life!" and someone behind you is saying, "I can't wait to get where that person is!"

- *Know that it's never too late.* If you feel that time has passed you by, you could be discouraged and feel that your best years, and all the dreams you dreamed, are a lost cause. Keep Ollie and Ben in mind: Ollie was sixty-eight years old when the military was finally desegregated, and Ben was eighty-six years old when he received his fourth star. If you've already tossed your dreams in the trash bin, snatch them out, dust them off, and get to work on making them a reality.

Own Your Path

Many people in the generations before us struggled and sacrificed in ways we may never understand or appreciate. They fought through and past every no so that our lives could be filled with yeses. They often chose to live an invisible life so that our lives could be visible. What better way to honor them than to use our visibility to make *them* visible?

No matter the circumstances, you can be a change agent in your family. Like my dad always told me: control what you can. You can choose to allow joy, optimism, and hope to become your modus operandi, to initiate a healing process that transforms you and your family. As with the million-dollar star, this is a long game, and maybe that *is* your million-dollar star: maybe it's not a literal monument or an amount of money—maybe it's overcoming to give yourself and those who follow you a healthier, happier future.

Never forget: *You are your descendants' dream ancestor.* Choose to live a life that will inspire future generations.

Acknowledgments

This journey of writing my first book was indeed the destination. To tell this story of my family—a story of American history—was itself a dream come true.

Discovering and uncovering this story has brought me together with incredible people from so many areas and walks of life. Whether it was a bread crumb to follow, a seed to plant, a lane to drive down, or a path to explore, support came in numerous ways—through time, access, trust, and appreciation.

When it comes to the arts, I gravitate toward the written word. Words are the foundation of a great story. As a verbal storyteller, I don't always write down my thoughts. Yet this book validated to me that the world is open to great stories being written.

I'd like to start by thanking and giving flowers to the incomparable team of people who worked with me, hand in hand and step-by-step, to make this book a reality.

To Amanda Bauch. You are amazing. You helped me understand how to bring a book alive, give it a voice, a heartbeat, and a soul—as well as the texture needed to turn 150 years of stories into a single work. This book would not have been possible without you. I thank the universe for bringing us together. You are a gift and an incredibly talented and humble person, who always went the extra mile.

To Charlamagne, your commitment to bring new voices to the forefront has been aligned with my vision since this book was just a story in my thoughts. You were the publishing team I wanted to work with the most—and you saw in me what I saw in you. To Karen Kinney, thank you for being so engaged and supportive from day one.

To my extraordinary editor, Nick Ciani. What can I say? You helped me understand the how and what of making words work, and their impact. Your guidance was invaluable to how this story should, could, and would be told. To Shida Carr, you have been a voice and an ally for me for years. Please know those efforts are felt. Many thanks to the Black Privilege Publishing team, from editorial and design to marketing and publicity staff, for all you've done to help make this book a reality: Hannah Frankel, Zakiya Jamal, and so many more.

To Emily Varga and Erin Simpson, for believing in the power of my voice to speak this story into the world.

To Nena Madonia Oshman, whom I met in 2008, along with Jan Miller of Dupree Miller & Associates. Collectively, you opened up my eyes to the world of publishing. To Nena and the Nominate Group, there are no words to express my thanks to you for taking a chance on a first-time author who didn't know the weight of what that meant. You have been my compass.

Ilana Glazer, you brought me into your circle of trust and helped me understand the power of family. Your warmth and voice opened a new world to me. Kelsie Kiley, I appreciate you for being a guiding light toward understanding storytelling.

To William Sidarweck, my right-hand man and business partner since high school. You are the most organized person I know, and your management of the process, the research, the paperwork, and the logistics were invaluable. Your contributions are felt daily.

To Leah Lakins, for spending weekends filled with Post-it Notes and helping me take hundreds of thoughts and organize them from ideas into the structure of a book.

To Rashida Peters, you are one of the most talented people I know. You helped educate me and guide me to visualize and imagine my thoughts in new and creative ways around this story. You brought them to life.

To Marzi and Sarah March of Bay PR. You have been there with me for the past ten years, along with Nikki Glor, to help me amplify and deliver my story to the world.

To Keith Major, one of the most talented photographers and creative directors in the world, for helping me bring out my best in front of a camera.

To the team who helped me behind the scenes. To Bill Britt, a wonderful ally and supporter of our family on so many levels. To Dave Kozlowski, your insight and input has been helpful to me both personally and professionally. To Don Manning, for reconnecting and putting in the time to reopen a piece of my family history.

To the West Point family. Thank you for welcoming our family back into yours in the most unbelievable way. To the West Point Davis Barrack dedication team, who spent countless hours to ensure our family legacy lives on forever. To Superintendent General Robert Caslen, your leadership was instrumental. To Colonel Rod Doyle, in some ways we bonded like brothers. Our paths crossed to create history. To First Lieutenant Simone Askew, you were the first captain at an important moment and made your voice heard. Thank you.

To Carmine Cocchia, you went the extra mile to shoot, edit, and produce a story of our family. Your creativity and craftsmanship with storytelling are award-winning. And your joy is contagious. Next up, that Emmy!

To Sherman Fleek, your presence was felt throughout the process of understanding the overall history and context of West Point and how my family played a part in its history.

To Archie Elam, for being so influential with the West Point AOG, and their support of all the Davis Barracks efforts. You guided the ship throughout the mission. To the CT West Point Association for giving me the stage to share this story with your community.

To Major Pat Locke, for being open and supportive of our family and for being an incredible leader and figure that commands the room as only you can.

To Superintendent General Darryl A. Williams and to Lisa Benitez, thank you for your continued support. To Jennifer Hicks-McGowen for inviting me back to listen, participate, and speak to the cadets.

To the US Army Corp of Engineers, Michael Embrich, and others for inviting me into the process of the construction and design of the Davis Barracks. Your team built a forever building for a forever story. I was proud to have met so many of them.

To the Tuskegee Airmen, Inc., for being encouraging and showing up to support your commander and friend General Davis whenever he is honored. The red jackets always bring out a feeling of inspiration in people.

To General Leon Johnson. Leon, you are a modern-day hero. You have shared so much with me about the history and legacy of the Air Force through your eyes and those of the Tuskegee Airmen. Your home is a museum to the Airmen. Your brain is the ChatGPT of Airmen information. Your support of our family has been unwavering.

To the USAFA—wow. You embody everything that General Davis stood for, leaned into, and believed.

To the Davis Airfield team, you have honored our family in a way that can only be stated from a place of humility and gratitude.

To Dr. Charles D. Dusch Jr., I appreciate you walking me through and documenting this moment in history and going the extra mile for the family. To Superintendent Lieutenant General Jay B. Silveria;

Secretary of the Air Force, the Honorable Barbara Barrett; former Chief of Staff of the Air Force, General David Goldfein; Chief Master Sergeant of the Air Force, Chief Master Sergeant Kaleth Wright; and Lieutenant Colonel Julian Stephens—you were the voices in the room that delivered on the Air Force Academy's value of "Excellence in All We Do."

To the chairman of the Joint Chiefs of Staff, General Charles "CQ" Brown Jr., you have made every effort to be present and show up to support and speak on behalf of the family. We are thankful for your time and thoughtfulness. General Davis would be so proud to see you at the levels you have climbed as a leader representing the US Air Force. You embody what he dreamed.

To the USAFA AOG. To Lieutenant General (Ret.) Mike Gould, CEO, you have continued to work behind the scenes to bring more visibility to the airfield and to our family. Words cannot express our gratitude. To Kelly Banet, for being helpful in continuing to build toward the future.

To the First Flight Society, you bestowed one of your highest honors upon our family. To William Douglas, for being a voice for aviation, and for recognizing and rewarding the accomplishments of those in the field of aviation with visibility and a moment of honor and respect. To Mike Fonseca, for leading with purpose to recognize the importance of December 17.

To the Smithsonian, you opened your doors to let me access pieces of my family that I'd never seen before and went the extra mile to make sure I got everything I needed. To Marilyn Graskowiak, for being so open to share your personal stories, and to Elizabeth C. Borja, not only for being a great host but also for cooking the recipes of Mrs. Agatha Davis!

To the Army War College, for giving me a passageway into a complicated past and being transparent with your time and resources.

To the people who inspired me through the days that seemed like marathons.

To Rob Schwartz, you believed in me from day one. Thank you for showing me how to be a creative leader and believing in my true voice as a disruptor. Your coaching is best in class, as is your friendship.

Tiffany R. Warren, you have shown me, through actions, how to live through purpose. You created a table, ADCOLOR, and invited me to have a seat. And you laid the pathway for corporate diversity that I still travel on today.

Elaine Welteroth, you inspired me to better uncover my own identity through writing. Your advice and vision of "more than enough" was the spark that lit my fire to get me here.

To my family . . .

To my mom and dad, sixty years married, same address, same home—as consistent as the wheel every step of the way. Thank you for teaching me and allowing me to dream big.

To Dad. This book is my love letter to you. You are my idol, role model, and mentor. Your humility is part of your genius. I am not me without you. You created me.

To Mom. My best friend. No words could ever be enough for the wonderful, immeasurable time, love, and support you have given me, beginning with life itself.

To my brother, Scotty, and sister, Sonja-Lisa. We are indeed the three-headed monster, different yet similar. Thank you for being my therapists, advisors, and confidants. To Scotty, thank you for teaching me so much, beginning with how to walk. And Sonja-Lisa, thank you for helping me become better linguistically with my writing. I'm inspired by your creativity.

To Angelique. Thank you for believing in me and being by my side on this journey. In some ways, Ben brought us together through our fathers.

The Burns Family. I was told if I forgot anyone I would be excommunicated. So many thanks to Trey Burns, Jill Burns, Madison Burns-Henry, Corbin Burns, Langston Burns, Ford Burns, Danielle Burns Wilson, Cy Wilson, and David Wilson.

To Stephanie Fornaris, Sarah Fornaris, Sophie Fornaris, and Lisa Scott. I love that we are family, and I appreciate every moment we've had together. To Sarah, you are my spirit animal, and I'm glad we got closer in NYC. To Melissa Scott Webster, for being a super-inspirational person and mother, and to Jeff Webster, Mia Branch, and Noah Webster.

To my second dad, Ron Douglas, aka Sings, Old Hollywood, and My . . . ! Cut from the same cloth, no words will ever be able to express my gratitude for your support in this life. Your love and guidance for me are immeasurable and incredible to experience. To my late uncle Bill (Dougie) Douglas, my namesake. I wish you could be here to see this come to life. You were my sounding board for this story. Our last meal together is still etched in my mind.

To my cousins Chante Hardesty, Matt Douglas, and Meghan Douglas—I appreciate you.

To my chosen family . . .

James Lesure, you have always been only a phone call away. You answer and give me that big brother advice, and you helped me understand the rules of engagement.

To the Hilfiger family. To Andy Hilfiger, you have helped me in so many ways, both near and far. You are incredible. To Kim Hilfiger, your smile is contagious. To Tommy, you helped change the course of my life with your insight and support. To Sheila Cox, twenty years of support and laughs—you have been an incredible ally. To Betsy Hilfiger, you are an angel on earth. You have been like a second mom to me. To DiDi, Ginny, and Bobby, your support and friendship have been so appreciated. To Mike Fredo and

Joe Fredo, you are two brothers who brought me into your family. Our bond inspires me.

To Norma Harris, your guidance, advice, and friendship with our family have been a gift. To Jay Brown, thanks for bringing me into your circle and helping me grow as a person and in business. To Kairi Brown, if I had another brother here on earth, it would be you. To Monica Brown and Dominique Brown, thank you for always treating me like family.

To Earvin "Magic" Johnson. Mr. J., you pulled me up in a way that forever changed my life. You are the mentor I always wanted and needed. I greatly appreciate what we accomplished and what you taught me.

To the Magic Johnson Enterprises family, Andre Johnson, for always being a great friend and colleague. To Kawanna Brown, for always being an incredible leader and ally. To Natalie Wilson, for always making time for me and helping connect the dots. To Mark Scoggins, thank you.

To Christine Simmons, for being a mentor from afar, unapologetically.

To Aneesha Salem, Tiffany Matthews, Amber Grant, and Sherria Heath, for always bringing grace to the space.

To Allison Kluger, for elevating and inspiring me. You walked me onto the Stanford campus and showed me what it was like to learn about reputation. Then you helped me develop my story to where it is today. Thank you.

Cal Hunter, you've taught me about so much. Thank you for your steadfast and unwavering support of me and your generosity of knowledge and time.

To John Cotton, you have been helping me show up looking put together. Always making time for me in the chair. Critical.

To JP Rizzitelli, you have been balancing the path for a decade. Your laugh is the best, and your patience with me is admirable.

To Lynne Hale for being so pivotal in helping reintroduce the Red Tail story to the world in 2020. Your effort and input to continue to ensure that the story of the Airmen lives on have been invaluable. I still smile thinking back to our dinner in Beverly Hills.

To Emory Jones, for creating my fave threads to rock, Paper Planes. I've got the crown in every color. Let's collab on a red Ben Davis edition.

To Alvin Bowles, the journey has been long. But no words can express my thanks for your mentorship and friendship.

To Jonathan Priester, for always being supportive and giving light to my vision.

To Claradith Landry, for connecting me and opening my eyes to new ideas and new rooms, and to the families of other military heroes who also paved the way: Henry Flipper, Charles Young, Jesse Leroy Brown, the 6888.

To Porter Braswell, you have taught me a lot about the economics of how things work. In many ways, your book inspired me to write mine. The best is yet to come.

To Aaron Walton, your moves across advertising and media have inspired me to tell this story. You've promoted storytelling in a way that allows people to see themselves in your work.

To Terry Albrighton, you have been a guiding light, offering perspective and principles. Keep shining.

To Sukari Pinnock, for helping me better understand my craft and for continuing to push me, as well as for educating a world of future leaders by inspiring them to push the bar forward.

To Lindsay Wagner, for being there to listen and collaborate on better ways of working and moving forward. Your input is valuable beyond measure.

To VFW–Benjamin O. Davis Post 311, for dedicating their VFW to Ben. To Bill Browne, for reaching out to me years ago and inviting me in.

To Benjamin O. Davis High School in Houston, Texas, and Benjamin O. Davis, Jr. Middle School in Compton, California, thank you for continuing to educate and inspire tomorrow's leaders. It's important you know the family appreciates all you do.

To the presidential libraries, for archiving historical moments that can be relived and reflected on, to inspire new generations.

In addition, I want to thank my circle, who provided and voiced their support along the way: Kerstin Emhoff, Mel Vargas, Justin Hillian, Wendell Haskins, Samir Sama, Mark Hamlin, Georgetta Foreman, Amy Twilley, Matt Rosenberg, Aaron Samuels, Alex Thomas, Ryan Ford, Stephen Burks, Iman Oubou, Eboni K. Williams, Corey Smith, Cynthia Maria Blanco, Chris Iki, Donna Graves, Josh Beckerman, Donnie Graves, Denis Strieff, Troy Ruhanan, Nancy Reyes, Monica Torres, Cynthia Clarke, Briana Mercado, Dr. Joi Spraggins, Emanuela Vargas, Russell Redeaux, Conan Milne, QDIII; my godfather the Honorable Lenny Cocco and his wife Jeanie Cocco, Tiffin Jernstedt, LeRoy Gillead II, Marcela Madera, Kim Nottingham Joyner, Tira Feierstein, Susie Nam, Marc Strachan, Bailey Pelkey, Maggie O'Brien, Keesha Jean Baptiste, Asha Davis, Alex Cuevas, Lana Alexio, Mickael Leroy, Andy Deaza, Adam Wallitt, Mr. Lee, Erica Lee, Jennifer Palacios, Renee Cafaro, Karl Carter, Maisha Dyson, Kristin Conte, Alissa Conte, Criseli Roddy, Jay Tibbits, Avan Hardwell, Josh Jacobson, Felicia Geiger, Meg Ireland Young, Val Greenberg, Josh Nackman, Harmonie Kreiger, Renee Airo-Ford, Vanessa Sandoval, Naheem Ballah, Robert Harwood-Matthews, John Cohen, Mark Clennon, Stephen Kim, Kyle Henning, Amber Grant, Natalie Robinson, Laura Davis, Anastasia Williams, Rhonda George, Kim Simun-Janson, Jesse Dunne, Kiley Dean, Toni Pringley, Khaki Stergio, Maria Kelts, Von Harris, Kevin Gigerure, Chauncey Plummer, Michael Tetuan, Alex Rojas, Courtney Friel, Justin Reyes, Marisol Vargas, Marcus Morales, Angela

Guzman, Max Rutherford, Saumya Vasuthevan, Manny Kess, Elena Kess, Omar Hamilton, Duff Stewart, Jordan Zimmerman, Adrienne Alexander, Christena J. Pyle, Kirya Francis, Andrea Beateu, Lisa Bashi, Cliff Courtney, Felita Harris, Jayanta Jenkins, Pam Gillingham, Lizzy Morgan, Kelly Williams, Shannon Guse, Stephen Hanson, Bryan Martos, Jason Barber, Iesha Reed, Luke Watson, Taylor Humphries, Josh Gross, Adu Adu, Constance Frazier, Ayanna Jackson, Joan Baker, Rudy Gaskins, Adrienne Chalfant Parker, Amber Guild, Phil Metz, Michael Tennant, Dr. Anyitsi Reynolds, Nicole Rocklin, Mario Armstrong, Julia Silverton, Alex Álvarez, Mary Nitollo, Carron Brown, Josh Newman, Doug Rowell, London Wright-Pegs, Erica Lovett, Talitha Watkins, Big Rob Feggans, Phil Burgess, Sabine Ellini, Arielle Winfield, Aki Spicer, Meghan Holston-Alexander, Jerico Cabayasa, Paige Blake, Stephen Medina, Derek Dolin, Mikkelangelo, Avan Hartwell, Sherry Riad, Billy Panagiotopoulos, David Saalfrank, Jason Taylor, Kris Friday, Kami Friday, S. Brooke White, Paulina Garcia, Lyndan Linnebank, Matt DeSiena, Idrees Dubar-Bernard, Stephen Burke, Pretty Til Dawn, Delphine Morisio, Joy Middleton-Saulny, Luke Watson, Tetiana Anderson, Anton Gunn, Amelia Azmi, Ward Niou, Brian Packin, CBH, Makisha Noel, Shane Santiago, Natalie Robinson, Pam Gillingham, Chad Carr, Alex Alvarez, NiRey Reynolds, Alejandro Rodriguez, and Ron Warner, among others. And to the Persaud brothers, for being such positive influences since 2001.

To my hometown friends in Bridgeport, Connecticut: the DiNardo family, Al Bauco, Ara India Hodge, Sal Murano, Joe Devellis, Sam Devellis, Jay Rodriquez, Ray Giasulo, Sandy Gonzolas, Priscilla Garcia, Thyhason Adams, Rob DeFilipo, Peter Vecchiarelli, Nancy Vecchiarelli, Lucy, and Big Pete. And to Miss Zembrzuski—I did it! I wrote the book just like we talked about in high school.

APPENDIX

Reflection Questions

Now that you know the story of the Invisible Generals and my quest to reclaim and preserve my family legacy, my hope is that you've been prompted to start having conversations with your relatives. I also hope you're feeling compelled to do some self-reflection, to better understand your thoughts and feelings about your own family legacy.

To that end, these reflection questions are intended to help you dig deeper into the ideas presented in this book and spark discussion with others.

Preface: A Man on a Mission

- Have you ever thought about how many hours you spend with your loved ones and don't know their story? Start having the conversation while you still have the opportunity.
- Can you imagine what your ancestors would think of you and your life? Would they be proud?
- If you could ask someone in your family any question about their past, who would that be, and what would you ask them?

Introduction: What's in a Name?

- Have you ever felt invisible or overlooked? If so, how did you handle that situation?
- What has prevented you from asking your elders their stories? Is it a traumatic experience that took place? If so, who benefits from the fear we've accepted around exploring our pasts?
- Could knowing more about your family history—both the good and the bad—uncover an inner passion or purpose that you may be seeking and drive you to the next level?

Chapter 1: Generational Collateral

- What kinds of generational collateral have you received that has provided you with opportunities?
- Do you have an example of a powerful story from your family that could challenge or inspire someone else? If so, share it with someone today!
- How are you living a life that will benefit the generations that come after you?

Chapter 2: Extraordinary Just to Be Ordinary

- In what ways is your life "ordinary" that previous generations might have considered extraordinary? What privileges do you enjoy that they didn't?
- What strategies can you use to avoid becoming complacent in your day-to-day life?
- Is there a person in your life whom you consider extraordinary? (If you can't identify someone in your personal life, think of an example from history.) In what ways is that person extraordinary, and what can you learn from their example?

Chapter 3: West Point's Invisible Alum

- Have you felt that your voice, or the voices of people you know, were marginalized? If so, how did that feel?
- How do you respond or react when you see people who look like you represented in different spheres, or when you hear voices you don't usually hear speaking in public spaces?
- What resources can you leverage to serve as a voice for the voiceless, to advocate for those who have no advocate?

Chapter 4: Living Life in a Liminal Space

- Who in your life has been the biggest inspiration? The most impactful mentor? The loudest cheerleader?
- Have you ever felt that your skills and abilities were undervalued? If so, how did you navigate that situation?
- In what ways are you staying prepared to seize opportunities to either discover or fulfill your purpose?

Chapter 5: The Pursuit of Double Victory

- Has an advocate ever appeared on the scene to help you move closer to achieving one of your goals? If so, who was that person, and how did they help you?
- Have you ever experienced being in the right place at the right time? If so, how did you find yourself in that situation, and what resulted?
- Are you a patient person? If you are, what situations and circumstances in your life created that trait, and how has it benefited you? If you don't have patience, what can you do to cultivate it, so you can play the long game?

Chapter 6: Founding Fathers

- Have your accomplishments ever been overlooked? If so, how did that make you feel, and how did you respond to the situation?
- Do you have a strong support system? Who's in that inner circle, and what do they contribute to your life? If you're lacking any kind of support, is there someone who could provide that, such as a coach, mentor, sponsor, or counselor/therapist?
- What are some of your favorite stories of perseverance, either from history or your family? How can you share with other people the lessons you've learned from these stories?

Chapter 7: Look Forward, Not Down

- Have you ever had to pivot in your life or ended up somewhere you didn't expect? If so, why did you need to make a change and how did things turn out in the end?
- In what ways are you currently serving others, whether in your home, workplace, or community? Are there other service opportunities you can explore? If the need seems too large for you to tackle on your own, who can you recruit to help you?
- How do you handle disappointments and setbacks, whether in your personal or professional life? Do you let it drag you down, or do you use it as a springboard for learning and growth?

Chapter 8: The Million-Dollar Star

- Do you discuss or get involved in politics? Why or why not?
- Have you or anyone else in your family ever had to wait a long time for recognition of an accomplishment, or to achieve a

major life goal? If so, what did that wait look like, and how did it feel when the recognition was received?

- Can you think of a "million-dollar star" in your family that unlocked opportunities for you and/or others (e.g., a college degree, a real estate purchase, etc.)? What benefits did that million-dollar star impart?

Chapter 9: Life after Death

- Have you experienced losing someone before you had a chance to ask them important questions or tell them something you wanted to say? What did you want to ask or say?
- Have you ever discovered a shocking family secret? If so, how did that affect you and/or your family?
- What is one thing you want to make sure your loved ones know about you and your life before your death?

Chapter 10: Monumental

- Have you ever given much thought to the monuments in your community? If so, what do you know about their history and what they represent? If not, how can you begin the learning process?
- If you believe that any of these monuments don't represent the best your community has to offer the world, what practical steps can you take to address the matter?
- Project fifty years into the future and think about what you want to see manifested in relation to your story. Do you want a monument erected? A scholarship fund established? A college course offered on the subject? What can you do today to start that process?

Chapter 11: The Brand Equity of Fairness

- When you think about brands you love, which of their characteristics most appeal to you? Alternately, if you've ever abandoned a brand, why do you no longer associate with it?
- How would you describe your personal brand? Or your family brand? If you struggle to articulate your thoughts, spend some time refining. Practice describing it to a few people you trust and who will provide honest, constructive feedback.
- What small thing can you do today to elevate your personal brand or your family brand?

Chapter 12: American

- If a museum housed a collection from your family's history, what items would be archived?
- Do your personal and professional lives feel integrated or disjointed? As you walk in your purpose, what are some ways you can bring those two spheres of your life into closer alignment?
- What do you want your descendants or future generations to say about you? Write or type a brief biography, as written by someone eulogizing you at the end of your life. Keep a copy of it to refer to when you feel like you're losing your way or unsure about your purpose, to reignite your passion and focus.

Notes

Introduction: What's in a Name?

1. Even though Ben was a colonel during the events covered in the movie, he eventually became a general.

Chapter 1: Generational Collateral

1 Benjamin O. Davis Jr., *Benjamin O. Davis, Jr.: American* (Washington, DC: Smithsonian Institution Press, 1991), 2.
2. Since this is my family's story, I've opted to refer to General Benjamin O. Davis Sr. by his family name, "Ollie," though I know it's more informal. This approach has the added benefit of making it easier to distinguish Ollie from his son, Ben.
3. Davis Jr., 2.
4. Davis Jr., 2.
5. In 1900, the Black illiteracy rate was approximately 45 percent, mainly because, under slavery, it had been illegal to teach them to read and write. See "120 Years of Literacy," National Assessment of Adult Literacy, accessed April 10, 2023, https://nces.ed.gov/naal/lit_history.asp.
6. Davis Jr., 2.
7. "Last Will and Testament of Louis P. H. Davis," Benjamin O. Davis Sr. Collection, Box 40, Folder 4, U.S. Army Military History Institute, Carlisle Barracks, Pennsylvania.

Chapter 2: Extraordinary Just to Be Ordinary

1. National Park Service, "Buffalo Soldiers," U.S. Department of the Interior, updated October 26, 2022, https://www.nps.gov/chyo/learn/historyculture/buffalo-soldiers.htm.
2. "Buffalo Soldiers of West Point," Buffalo Soldiers Association of West Point, accessed September 1, 2022, https://www.buffalosoldiersofwestpoint.org/buffalo-soldiers-at-westpoint; Michael Davis and Miriam Kleiman, "National Archives Unveils Photos of Buffalo Soldiers at West Point," National Archives, February 26, 2021, https://www.archives.gov/news/articles/archives-unveils-photos-of-buffalo-soldiers-west-point.
3. Marvin E. Fletcher, *America's First Black General: Benjamin O. Davis, Sr., 1880–1970* (Lawrence: University Press of Kansas, 1989), 9.
4. "Correspondence from M. B. Hughes Regarding Benjamin O. Davis," Benjamin O. Davis Sr. Collection, Box 7, Folder 1, U.S. Army Military History Institute, Carlisle Barracks, Pennsylvania.
5. "Pres. Clinton's Remarks Honoring Gen. Benjamin O. Davis, Jr. of the Tuskegee Airmen," clintonlibrary42 video, YouTube, April 4, 2013, 34:12, https://www.youtube.com/watch?v=rIijkuGjtnI.
6. Benjamin O. Davis Jr., *Benjamin O. Davis, Jr.: American* (Washington, DC: Smithsonian Institution Press, 1991), 3.
7. "Correspondence from M. B. Hughes."
8. Davis Jr., 12.
9. "Memorandum for the Chief of Staff Regarding Employment of Negro Man Power in War, November 10, 1925," President's Official Files 4245-G: Office of Production Management: Commission on Fair Employment Practices: War Department, 1943, US Army War College; Archives of the Franklin D. Roosevelt Library, 5, accessed October 24, 2022, https://www.fdrlibrary.org/documents/356632/390886/tusk_doc_a.pdf/4693156a-8844-4361-ae17-03407e7a3dee.
10. Davis Jr., 25.

11. "Benjamin Oliver Davis, Sr.," U.S. Army Center of Military History, updated January 31, 2021, https://history.army.mil/html/topics/afam/davis.html.

Chapter 3: West Point's Invisible Alum

1. Benjamin O. Davis Jr., *Benjamin O. Davis, Jr.: American* (Washington, DC: Smithsonian Institution Press, 1991), 14.
2. Davis Jr., 14.
3. Davis Jr., 19–20.
4. Davis Jr., 20.
5. Davis Jr., 22.
6. Davis Jr., 22.
7. James Feron, "Blacks in a Long Gray Line," *New York Times*, June 2, 1991, https://www.nytimes.com/1991/06/02/us/blacks-in-a-long-gray-line.html.
8. Davis Jr., 27, 28.
9. W. E. B. Du Bois, "Charles Young," *The Crisis* 23, no. 4 (February 1922): https://archive.org/details/sim_crisis_1922-02_23_4/page/154/mode/2up.
10. Davis Jr., 29.
11. "General Benjamin Oliver Davis Jr.," Military Hall of Honor, accessed March 31, 2022, https://militaryhallofhonor.com/honoree-record.php?id=701.
12. MMS, "African-American Soldiers in World War I: The 92nd and 93rd Divisions," EDSITEment!, National Endowment for the Humanities, November 13, 2009, https://edsitement.neh.gov/lesson-plans/african-american-soldiers-world-war-i-92nd-and-93rd-divisions.
13. "No. 1 Graduate of the Nation," *The Crisis*, August 1936, Benjamin O. Davis Jr. Collection–Graduation, Smithsonian National Air and Space Museum Archives, https://transcription.si.edu/view/15930/NASM-NASM.1992.0023-M0000045-00200.
14. Davis Jr., 53.

Chapter 4: Living Life in a Liminal Space

1. Benjamin O. Davis Jr., *Benjamin O. Davis, Jr.: American* (Washington, DC: Smithsonian Institution Press, 1991), 8–9.
2. Davis Jr., 67. Ben's assertion is supported by news reports from the time, such as Morgan Michaels, "The Private War of General Davis," *Tropic*, February 23, 1969.

Chapter 5: The Pursuit of Double Victory

1. Cassie Peterson, "The History of the Civilian Pilot Training Program," *Plane & Pilot*, updated June 3, 2022, https://www.planeandpilotmag.com/news/pilot-talk/2022/06/03/plane-facts-civilian-pilot-program/.
2. Sherri L. Smith, *Who Were the Tuskegee Airmen?* (New York: Penguin Workshop, 2020), 17.
3. Tracie Reddick, "Tuskegee Airman Yenwith Whitney Soared Above Barriers," *Bradenton Herald*, July 27, 2000, http://www.bradenton.com/news/article34510047.html.
4. Benjamin O. Davis Jr., *Benjamin O. Davis, Jr.: American* (Washington, DC: Smithsonian Institution Press, 1991), 70.
5. Smith, 39.
6. "99th Pursuit Squadron Inaugurated," *Newspic*, August 1941, in George L. Washington, *The History of Military and Civilian Pilot Training of Negroes at Tuskegee, Alabama: 1939–1945*, 1972, accessed December 30, 2022, http://archive.tuskegee.edu/repository/digital-collection/tuskegee-airmen/history-of-military-and-civilian-pilot-training-of-negroes-at-tuskegee-1939-1945/.
7. Davis Jr., 69.
8. "Tuskegee Airmen," History.com, updated January 26, 2021, https://www.history.com/topics/world-war-ii/tuskegee-airmen.
9. Davis Jr., 81.

10. Rhonda Crowder, "Black Newspapers Have Long Recorded Our History," Cleveland.com, updated February 24, 2010, https://www.cleveland.com/call-and-post/2010/02/black_newspapers_have_long_rec.html. In the twenty-first century, Black newspapers have met the same fate as many periodicals, and only about two hundred are still around. However, numerous online Black news outlets continue to flourish.
11. Matthew Delmont, "Why African-American Soldiers Saw World War II as a Two-Front Battle," *Smithsonian Magazine*, August 24, 2017, https://www.smithsonianmag.com/history/why-african-american-soldiers-saw-world-war-ii-two-front-battle-180964616/#:~:text=The%20Double%20Victory%20campaign%2C%20launched%20by%20the%20Courier,Nazism%20and%20white%20supremacy%20in%20the%20United%20States.
12. Douglas Melville, "General Benjamin O. Davis, Jr. (Commander of the Tuskegee Airmen)—Funeral Ceremony—July 4th, 2002," YouTube video, December 5, 2019, 30:20 to 36:49, https://www.youtube.com/watch?v=XfbKsuSDR5M.
13. Davis Jr., 124.
14. Stephanie Siek, "Women of Tuskegee Supported Famed Black Pilots," CNN, February 2, 2012, https://www.cnn.com/2012/02/02/us/women-of-tuskegee-supported-famed-black-pilots/index.html.
15. Siek, "Women of Tuskegee Supported Famed Black Pilots."

Chapter 6: Founding Fathers

1. Benjamin O. Davis Jr., *Benjamin O. Davis, Jr.: American* (Washington, DC: Smithsonian Institution Press, 1991), 18–19.
2. "Harry S. Truman and Civil Rights," Harry S. Truman Library and Museum, accessed March 31, 2023, https://www.trumanlibrary.gov/education/presidential-inquiries/harry-s-truman-and-civil-rights.
3. Davis Jr., 155.

4. Davis Jr., 164.
5. Davis Jr., 165.
6. John L. Frisbee, *Makers of the United States Air Force*, rev ed. (1987; Washington, DC: Office of Air Force History, 1996), 230.
7. Please note that this wasn't the official TOPGUN Navy program popularized in the movies starring Tom Cruise. That program started in 1969. For more information, see Commander David "Bio" Baranek, "TOPGUN: The Navy's First Center of Excellence," U.S. Naval Institute, September 2019, https://www.usni.org/magazines/proceedings/2019/september/topgun-navys-first-center-excellence.
8. AARP, "The Untold Story of the First Top Gun Competition," YouTube video, 9:52, February 11, 2022, https://www.youtube.com/watch?v=H8QOp2mxUGA. Also see American Veterans Center, "'Tuskegee Top Gun' James Harvey, the First African American Jet Combat Pilot," YouTube video, 37:22, May 19, 2020, https://www.youtube.com/watch?v=VqA1ihi_0MU.
9. AARP, "The Untold Story."
10. Christine Fernando, "'Finally': Tuskegee Airmen Honored 73 Years After Competition Win Was 'Swept Under the Rug,'" *USA Today*, February 11, 2022, https://www.usatoday.com/story/news/nation/2022/02/11/tuskegee-airmen-honored-73-years-after-winning-air-force-contest/6718483001/.
11. Fernando, "'Finally.'"

Chapter 7: Look Forward, Not Down

1. Elizabeth Blair, "'Segregated Skies' Tells the Story of the First Black Pilot for a Commercial Airline," National Public Radio, February 18, 2022, https://www.npr.org/2022/02/18/1080731249/segregated-skies-tells-the-story-of-the-first-black-pilot-for-a-commercial-airli.
2. Mark Finlay, "David E Harris: The First African-American Pilot to Fly for a Major US Passenger Airline," Simple Flying, October 3, 2022, https://

simpleflying.com/david-harris-first-african-american-passenger-airline-pilot-story/.

3. Benjamin O. Davis Jr., *Benjamin O. Davis, Jr.: American* (Washington, DC: Smithsonian Institution Press, 1991), 331.
4. Davis Jr., 345.
5. Davis Jr., 362.
6. Weird History, "What Happened During the Golden Age of Highjackings," YouTube video, May 9, 2021, 12:02, https://www.youtube.com/watch?v=CCbyST-A238.
7. Davis Jr., 359.
8. Davis Jr., 385.
9. Brendan I. Koerner, *The Skies Belong to Us: Love and Terror in the Golden Age of Hijacking* (New York: Broadway Books, 2013), 76.
10. Davis Jr., 330.
11. Davis Jr., 379.
12. Secretary of the Air Force Public Affairs, "First Air Force Female Four-Star General Confirmed," United States Air Force, March 28, 2012, https://www.af.mil/News/Article-Display/Article/111462/first-air-force-female-four-star-general-confirmed/.
13. "Shawna Kimbrell: First African American Female Fighter Pilot," *Aerotech News*, January 31, 2020, updated February 2, 2020, https://www.aerotechnews.com/nellisafb/2020/01/31/shawna-kimbrell-first-african-american-female-fighter-pilot/.

Chapter 8: The Million-Dollar Star

1. "H.R.1119–105th Congress (1997–1998): National Defense Authorization Act for Fiscal Year 1998," Congress.gov, November 18, 1997, https://www.congress.gov/bill/105th-congress/house-bill/1119.

2. "Pres. Clinton's Remarks Honoring Gen. Benjamin O. Davis, Jr. of the Tuskegee Airmen," clintonlibrary42 video, YouTube, April 4, 2013, 28:14 to 28:32, https://www.youtube.com/watch?v=rIijkuGjtnI.
3. "Pres. Clinton's Remarks Honoring Gen. Benjamin O. Davis, Jr. of the Tuskegee Airmen," clintonlibrary42 video, YouTube, April 4, 2013, 34:12, https://www.youtube.com/watch?v=rIijkuGjtnI.
4. "US Air Force Retirement Pay Calculator," *Aircraft Air Force* blog, February 25, 2021, https://aircraftairforce.blogspot.com/2021/02/us-air-force-retirement-pay-calculator.html.

Chapter 9: Life after Death

1. "History of Arlington National Cemetery," Arlington National Cemetery, accessed January 20, 2023, https://www.arlingtoncemetery.mil/Explore/History-of-Arlington-National-Cemetery.
2. "History of Arlington National Cemetery."
3. Emily Goodman, "22 Things You Never Knew About Arlington National Cemetery," *Reader's Digest*, last updated November 28, 2022, https://www.rd.com/list/arlington-national-cemetery-facts/.
4. Douglas Melville, "General Benjamin O. Davis, Jr. (Commander of the Tuskegee Airmen)—Funeral Ceremony—July 4th, 2002," YouTube video, December 5, 2019, 22:40 to 41:40, https://www.youtube.com/watch?v=XfbKsuSDR5M.
5. "The U.S. Public Health Service Syphilis Study at Tuskegee," Centers for Disease Control and Prevention, April 22, 2021, https://www.cdc.gov/tuskegee/timeline.htm.
6. Elizabeth Nix, "Tuskegee Experiment: The Infamous Syphilis Study," History.com, updated December 15, 2020, https://www.history.com/news/the-infamous-40-year-tuskegee-study.

7. Sanjana Manjeshwar, "America's Forgotten History of Forced Sterilization," *Berkeley Political Review*, November 4, 2020, https://bpr.berkeley.edu/2020/11/04/americas-forgotten-history-of-forced-sterilization/; "The Supreme Court Ruling That Led to 70,000 Forced Sterilizations," National Public Radio, March 7, 2016, https://www.npr.org/sections/health-shots/2016/03/07/469478098/the-supreme-court-ruling-that-led-to-70-000-forced-sterilizations.

Chapter 10: Monumental

1. Jimmy Byrn and Gabe Royal, "What Should West Point Do About Its Robert E. Lee Problem?," Modern War Institute at West Point, June 22, 2020, https://mwi.usma.edu/west-point-robert-e-lee-problem/.
2. Simon Romero, "'The Lees Are Complex': Descendants Grapple with a Rebel General's Legacy," *New York Times*, August 22, 2017, https://www.nytimes.com/2017/08/22/us/lee-family-confederate-monuments-legacy.html.
3. Byrn and Royal, "What Should West Point Do About Its Robert E. Lee Problem?"
4. Benjamin O. Davis Jr., *Benjamin O. Davis, Jr.: American* (Washington, DC: Smithsonian Institution Press, 1991), 423.
5. "General Benjamin O. Davis Jr. Barracks," Coldspring, accessed June 21, 2022, https://www.coldspringusa.com/case-studies/davis-barracks/.
6. Psalm 118:22 (NKJV).
7. Michelle Eberhart, "Davis Barracks Time Capsule Cemented into Wall for 100 Years," U.S. Army, May 1, 2017, https://www.army.mil/article/187055/davis_barracks_time_capsule_cemented_into_wall_for_100_years.
8. Claudette Roulo, "Academy Names Airfield in Honor of First Black Air Force General," U.S. Department of Defense, November 1, 2019, https://

www.defense.gov/News/News-Stories/Article/Article/2006391/academy-names-airfield-in-honor-of-first-black-air-force-general/.

9. "USAFA Airfield Gets a New Name," Association of Graduates, United States Air Force Academy and the Air Force Academy Foundation, accessed December 26, 2022, https://www.usafa.org/News/DavisAirfield.
10. Serena Bettis, "FE Warren Air Force Base Home Dedication Memorializes First Black General," *Wyoming Tribune Eagle*, August 7, 2022, https://www.wyomingnews.com/news/f-e-warren-afb-home-dedication-decades-in-the-making-memorializes-a-buffalo-soldier/article_802f8016-71d5-5db5-af1b-949bd30c7a7b.html#:~:text=The%20Buffalo%20Soldiers%20of%20the%20American%20West%20present,American%20one-star%20general%20in%20the%20U.S.%20Armed%20Forces.

Chapter 11: The Brand Equity of Fairness

1. "Text—S.2052—116th Congress (2019–2020): A Bill to Authorize the Honorary Promotion of Colonel Charles E. McGee to Brigadier General in the United States Air Force," Congress.gov, August 2, 2019, https://www.congress.gov/bill/116th-congress/senate-bill/2052/text.
2. Dontavian Harrison, "Col. Charles Young Posthumously Promoted to Brigadier General at West Point," U.S. Army, May 2, 2022, https://www.army.mil/article/256278/col_charles_young_posthumously_promoted_to_brigadier_general_at_west_point.
3. "Tuskegee Airmen National Historic Site Quarter," United States Mint, last updated August 2, 2019, https://www.usmint.gov/coins/coin-medal-programs/america-the-beautiful-quarters/tuskegee-airmen-national-historic-site.
4. Reelblack One, "Benjamin O. Davis, Jr. Interview (1970)," YouTube video, November 11, 2018, 3:36, https://www.youtube.com/watch?v=xa8LWkkuI60.

5. Tiare Dunlap, "Veterans from World War II's All-Black Aviation Unit on How They Shattered Racial Stereotypes: 'We Proved That Thinking Wrong,'" *People*, April 20, 2016, https://people.com/celebrity/tuskegee-airmen-share-how-they-shattered-racial-stereotypes-in-world-war-ii/.
6. *The Daily Show with Jon Stewart*, season 17, episode 40, "George Lucas," directed by Chuck O'Neil, written by Rory Albanese et al., featuring John Stewart and George Lucas, aired January 9, 2012, on Comedy Central, https://www.cc.com/video/j67j6n/the-daily-show-with-jon-stewart-george-lucas.
7. *The Daily Show with Jon Stewart.*
8. The Black History Month commercial ended up airing during the Grammy Awards instead of the Super Bowl, but was still viewed online millions of times. You can watch the video on YouTube: https://www.youtube.com/watch?v=Qyiautg41h8.

Chapter 12: American

1. "Benjamin O. Davis Sr.: An Inventory of His Collection," Benjamin O. Davis Sr. Collection, the U.S. Army Military History Institute, Carlisle Barracks, Pennsylvania, accessed October 21, 2022, https://arena.usahec.org/.
2. Elizabeth Borja, "Benjamin O. Davis's Thanksgiving Turkey in Taipei," Smithsonian National Air and Space Museum, November 13, 2017, https://airandspace.si.edu/stories/editorial/benjamin-davis-thanksgiving-turkey-taipei; Elizabeth Borja and Melissa A. N. Keiser, "How Many Quarts of Tomato Soup?! Modern Takes on Historical Recipes from the Benjamin O. Davis Collection," Smithsonian National Air and Space Museum, July 7, 2021, https://airandspace.si.edu/stories/editorial/how-many-quarts-tomato-soup-modern-takes-historical-recipes-benjamin-o-davis.

3. Morgan Michaels, "The Private War of General Davis," *Tropic*, February 23, 1969. In Benjamin O. Davis, Jr. Collection, Acc. 1992.0023, National Air and Space Museum, Smithsonian Institution, accessed March 31, 2023, https://edan.si.edu/slideshow/viewer/?eadrefid=NASM.1992.0023_ref3505.
4. Chiayi Ho, "Median Line Issue Raises Questions over Beijing's Agenda," *Taiwan Today*, July 17, 2009, https://taiwantoday.tw/news.php?unit=10,23,45,10&post=15508.
5. Jesse Johnson, "'New Normal' as Chinese Warplanes Push across Taiwan Strait Median Line," *Japan Times*, September 7, 2022, https://www.japantimes.co.jp/news/2022/09/07/asia-pacific/china-taiwan-median-line/.
6. "Risk Map 2023 Analysis: Taiwan Strait," Global Guardian, October 10, 2022, https://www.globalguardian.com/newsroom/risk-map-taiwan-strait.

Epilogue: Inspired to Greatness

1. "Flying Machine Soars Three Miles in Teeth of High Wind over Sand Hills and Waves on Carolina Coast," *Virginian-Pilot*, December 18, 1903, Wright Brothers Aeroplane Company, https://wright-brothers.org/History_Wing/Wright_Story/Inventing_the_Airplane/December_17_1903/Virginia_Pilot_Story.htm.
2. Outer Banks Forever, "119th Anniversary of the First Flight at Wright Brothers National Memorial," Facebook, December 17, 2022, https://www.facebook.com/obxforever/videos/812654606473697.

Index

About the Author

DOUG MELVILLE is one of the most innovative voices in DEI, with more than a decade of experience. Most recently he was the global head of diversity and inclusion at Richemont, one of the world's leading luxury goods companies. There he sat on the governance and sustainability committee and led the effort to create policies and practices related to corporate DEI worldwide. This included a two-year project to ensure global pay equity across forty thousand colleagues, traveling to forty countries to workshop DEI, and teaching global international compliance.

Previously Melville worked on Madison Avenue to advise clients on inclusive ad campaigns and communication strategies, including with Apple, Amazon, Airbnb, PepsiCo, Nissan, and others. He also sat on the executive team of Magic Johnson Enterprises.

Melville's first book, *Invisible Generals* (Black Privilege Publishing), was released on Veterans Day 2023. He's a lecturer at Stanford University on reputation management and at Georgetown University's School of Continuing Studies DEI program. He has also worked with Harvard's negotiation and mediation clinical program to empower their top law students on the importance of DEI in corporate settings.